Phosphoplipid technology and applications

Also in the Oily Press Lipid Library:

Volume 21. Long-Chain Omega-3 Specialty Oils
Edited by Harald Breivik

Volume 20. Antioxidants in Food and Biology: Facts and Fiction
Written by Edwin N. Frankel

Volume 19. Lipids: Structure, Physical Properties and Functionality
Written by Kåre Larsson, Peter Quinn, Kiyotaka Sato and Fredrik Tiberg

Volume 18. Lipid Oxidation (second edition)
Written by Edwin N. Frankel

Volume 17. Bioactive Lipids
Edited by Anna Nicolaou and George Kokotos

Volume 16. Advances in Lipid Methodology – Five
Edited by Richard O. Adlof

Volume 15. Lipid Analysis (third edition)
Written by William W. Christie

Volume 14. Confectionery Fats Handbook
Written by Ralph E. Timms

Volume 13. Lipids for Functional Foods and Nutraceuticals
Edited by Frank D. Gunstone

Volume 12. Lipid Glossary 2
Written by Frank D. Gunstone and Bengt G. Herslöf

Volume 11. Lipids in Nutrition and Health: A Reappraisal
Written by Michael I. Gurr

Volume 9. *Trans* Fatty Acids in Human Nutrition
Edited by Jean Louis Sébédio and William W. Christie

Volume 8. Advances in Lipid Methodology – Four
Edited by William W. Christie

Volume 7. Advances in Lipid Methodology – Three
Edited by William W. Christie

Volumes 1– 6 and 10. Out of print

Woodhead Publishing in Food Science, Technology and Nutrition

Phospholipid technology and applications

Edited by

FRANK D. GUNSTONE
St Andrews, Scotland

THE OILY PRESS

WP
WOODHEAD PUBLISHING

Oxford Cambridge Philadelphia New Delhi

Published by Woodhead Publishing Limited,
80 High Street, Sawston, Cambridge CB22 3HJ, UK
www.woodheadpublishing.com
www.woodheadpublishingonline.com

Woodhead Publishing, 1518 Walnut Street, Suite 1100, Philadelphia,
PA 19102-3406, USA

Woodhead Publishing India Private Limited, G-2, Vardaan House, 7/28 Ansari Road,
Daryaganj, New Delhi – 110002, India
www.woodheadpublishingindia.com

First published by The Oily Press, 2008
Reprinted by Woodhead Publishing Limited, 2012

British Library Cataloguing in Publication Data
A catalogue record for this book is available from the British Library

ISBN 978-0-9552512-2-1 (print)
ISBN 978-0-85709-788-0 (online)

This book is **Volume 22** in **The Oily Press Lipid Library**

Typeset by Ann Buchan (Typesetters), Middlesex, UK
Printed by Lightning Source

Preface

Phospholipids are an important group of lipids with major areas of interest in biology, biochemistry and medicine, and also in science and technology. It is the latter that are emphasized in this book. The major source of phospholipids is the lecithin recovered during degumming of vegetable oils, particularly soybean oil. This crude material finds uses in its own right, but it can be purified through a series of processes that eventually lead to individual phospholipid classes such as the phosphatidylcholines.

It is widely accepted that oil and water do not mix, but there are several areas in science and technology where these two distinct phases must coexist in stable emulsions. This is achieved by admixture of amphiphilic molecules of which the phospholipids are important natural examples produced commercially at levels in excess of 250,000 tonnes per year. This property has been known for a long time by cooks who, while knowing little about lipid structure or amphiphilic systems, nevertheless recognized the importance of eggs (a rich source of phospholipids) in their recipes.

Today phospholipids find many uses in the food industry and in other industries that exploit the amphiphilic nature of these compounds. Further, there are now important procedures by which the amphiphilicity can be optimized for different uses.

The early chapters in this book are devoted to the more common glycerol-based phospholipids and cover their structure, source, composition, modification by chemical and enzymatic methods, their physical, chemical, and nutritional properties, and their major uses. The final chapter is devoted to another kind of phospholipid – the sphingolipids – in which there is a growing interest.

With its emphasis on science and technology, this book should be of special value to those in the food, cosmetic, and pharmaceutical industries.

I thank all those who have contributed to this book, especially Michael Schneider who collaborated with the editor and publisher in its early stages in developing the topics to be covered and suggested persons, some of whom became part of the writing team.

All the authors join with the editor in thanking Frances Daniel who has used her experience in copy editing to improve the appearance of these eight chapters and make them into a useful book.

Frank Gunstone
St Andrews
Scotland
March 2008

Contents

List of Contributors

Ashish Kumar Acharya, BioCentrum-DTU, Technical University of Denmark, Building 222, DK-2800 Kgs. Lyngby, Denmark

Alice Bonekamp, Technical Director (Lecithin), ADM Specialty Ingredients (Europe) BV, Seehafenstr. 24, D-21079 Hamburg, Germany

Anders Carlsson, MediGelium AB, S:t Göransgatan 80, SE-112 38 Stockholm, Sweden

Asim K. Duttaroy, Department of Nutrition, Institute for Basic Medical Sciences, Faculty of Medicine, University of Oslo, POB 1046 Blindern, N-0316 Oslo, Norway

Frank D. Gunstone, 3 Dempster Court, St Andrews, Fife KY16 9EU, Scotland

Zheng Guo, Assistant Professor, BioCentrum-DTU, Technical University of Denmark, Building 222, DK-2800 Kgs. Lyngby, Denmark

Ernesto Hernandez, Director, Process Development, OmegaPure Technology and Innovation Center, 6961 Brookhollow W Drive, Ste. 190, Houston, Texas 77040, USA

Åke Nilsson, Lund University, SE-221 85 Lund, Sweden

Nathalie Quezada, Texas A&M University, College Station, Texas, USA

Michael Schneider, LECITHOS – Functional Lipid Innovation and Consulting, Dr. Kausch-Str. 26, D-67251 Freinsheim, Germany

Anders Falk Vikbjerg, Research Scientist, LiPlasome Pharma A/S, Technical University of Denmark, Building 207, DK-2800 Kgs. Lyngby, Denmark

Tong Wang, Associate Professor, Department of Food Science and Human Nutrition, Iowa State University, 2312 Food Sciences Bldg, Ames, IA 50011-0001, USA

Xuebing Xu, Associate Professor, BioCentrum-DTU, Technical University of Denmark, Building 227, DK-2800 Kgs. Lyngby, Denmark. Current address: Professor in Agro-Biotechnology Science, Department of Molecular Biology, University of Aarhus, DK-8000 Aarhus, Denmark

Long Zhang, BioCentrum-DTU, Technical University of Denmark, Building 222, DK-2800 Kgs. Lyngby, Denmark

CHAPTER 1

Chemical structure and biological function

TONG WANG

Iowa State University, Ames, Iowa, USA

This chapter provides an overview of the chemical structure and biological functions of various classes of phospholipids. Processing, physical properties and chemical modifications are detailed in the following chapters in the book. Some clinical and nutritional properties of phospholipids are also reviewed in separate chapters.

A. Phospholipid molecular structure

Phospholipids (PLs) are a class of lipids that contains phosphorus in their molecules. Most often they are esters of phosphoric acid (H_3PO_4), but sometimes of phosphonic acid (H_3PO_3). The two basic classes of PLs are glycerol phospholipids and sphingomyelins. The structures of major plant and animal PLs are presented in Figure 1.1.

In glycerol PLs, the hydroxyl groups at the *sn*-1 and *sn*-2 positions of the glycerol are esterified with two fatty acids, and the hydroxyl group at the *sn*-3 position is esterified with phosphoric acid. The second free hydroxyl group of the phosphate ester (phosphatidic acid, PA) can react with other alcohols (serine, choline, ethanolamine and inositol) to form phosphatidylserine (PS), phosphatidylcholine (PC), phosphatidylethanolamine (PE), and phosphatidylinositol (PI). Phosphatidylglycerol (PG) and diphosphatidylglycerol (DPG, or cardiolipin) involve a second molecule of glycerol (see Figure 1.1). Sphingomyelins (SM) consist of sphingosine (a dihydroxy amine) bonded to one fatty acid at the NH_2 function and one polar head group such as phosphocholine or phosphoethanolamine at the CH_2OH function.

Phosphonolipids differ from PLs in that the phosphorus–carbon (P–C) bond links the aminoethyl group to phosphonic acid which is esterified with either a

Phosphatidylcholine (PC)

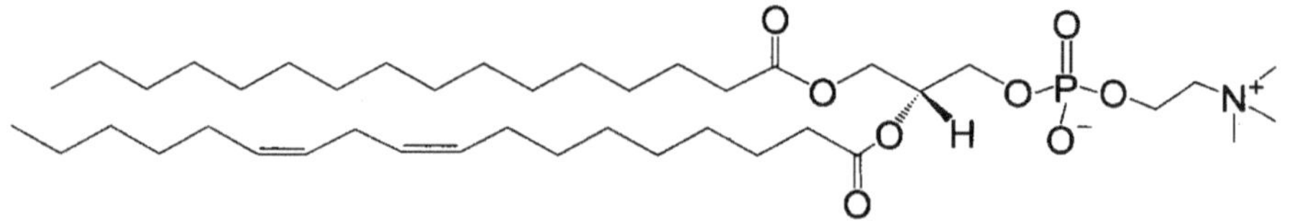

Phosphatidylethanolamine (PE)

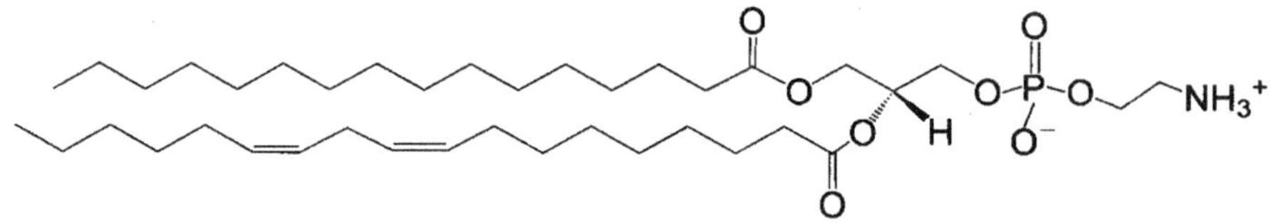

Phosphatidylserine (PS), sodium salt

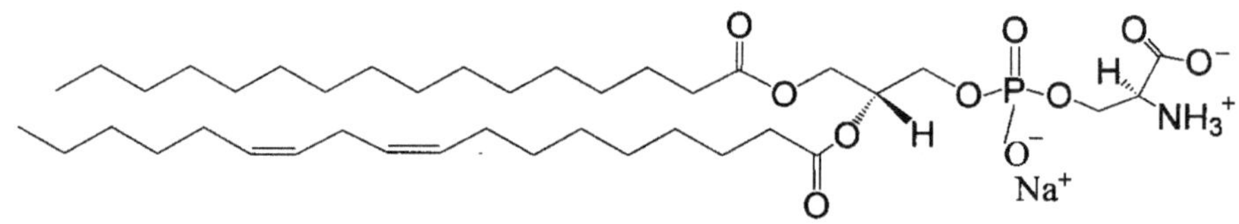

Phosphatidylinositol (PI), sodium salt

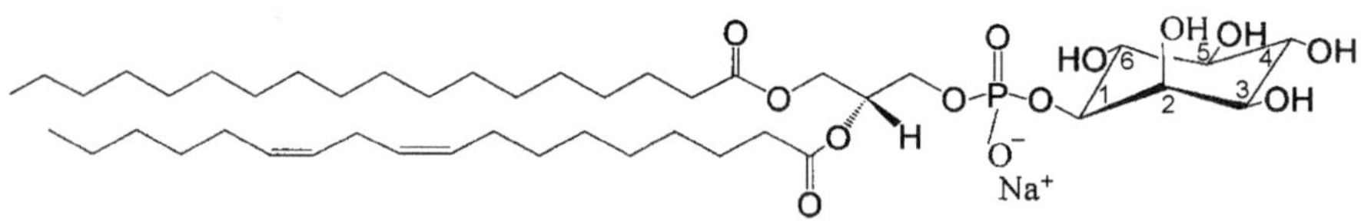

Phosphatidic acid (PA), sodium salt

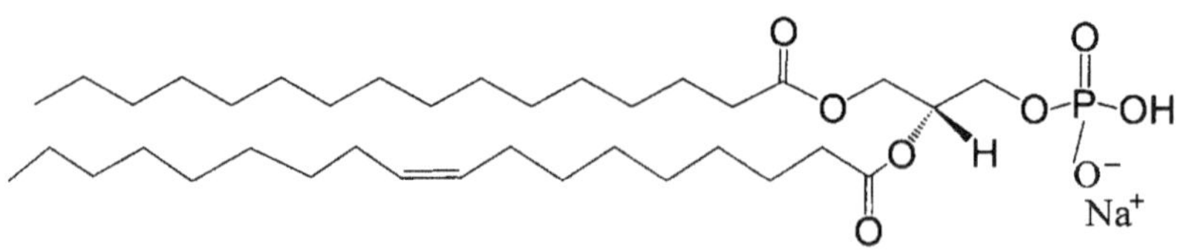

Phosphatidylglycerol (PG), sodium salt

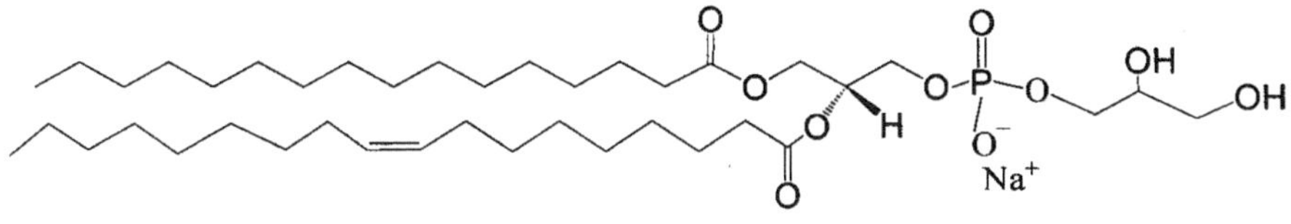

Diphosphatidylglycerol (DPG, or cardiolipin), sodium salt

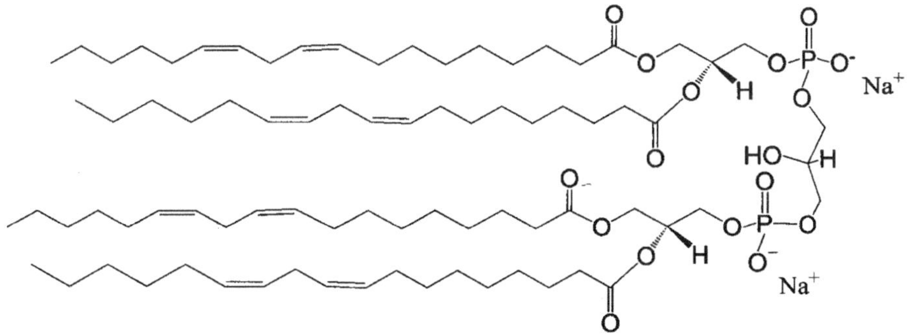

Sphingomyelin (SM)

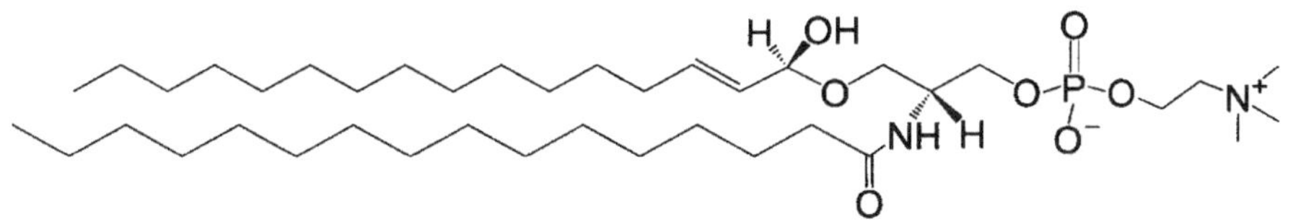

Phosphonolipids (1,2-diacyl- *sn*-glycerol-3-(2'-aminoethyl) phosphonate)

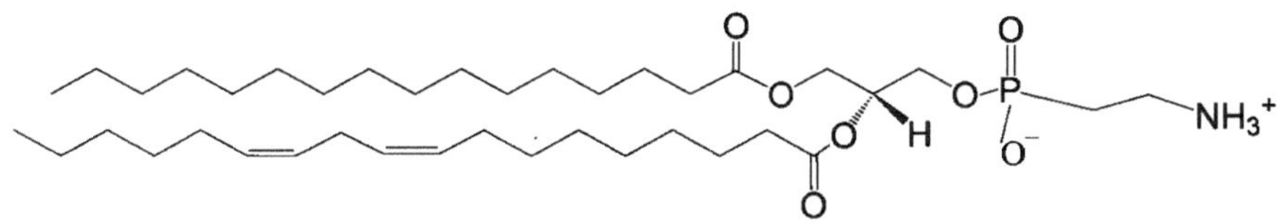

Figure 1.1. General structure of common phospholipids of plant and animal sources.

diacylglycerol or a ceramide moiety. A particular phosphonolipid that is an analogue of PE, 1,2-diacyl-*sn*-glycerol-3-(2'-aminoethyl) phosphonate, is shown in Figure 1.1.

In general, PLs can be synthesized by two mechanisms in biological systems. One utilizes a CDP (cytidine diphosphate)-activated polar head group to attach to the phosphate of PA, and the other utilizes CDP-activated 1,2-diacylglycerol to attach to the polar head group.

For PC synthesis, choline is first activated by phosphorylation and then by coupling to CDP prior to attachment to PA. PC is also synthesized by the addition of choline to CDP-activated 1,2-diacylglycerol. A third pathway for PC synthesis involves the conversion from either PS or PE. The conversion of PS to PC requires decarboxylation of PS to yield PE; this then undergoes a series of three methylation reactions utilizing *S*-adenosylmethionine (SAM) as methyl group donor. PE can be synthesized similarly to PC, and also by the decarboxylation of PS.

PS and PI are formed by a phosphoester linkage between the hydroxyl of an alcohol (serine or inositol) and CDP-diacylglycerol. PI can subsequently undergo a series of phosphorylations of the hydroxyl groups on the inositol ring to produce polyphosphoinositols. One of these, phosphatidylinositol 4,5-bisphosphate, PIP_2, is an important membrane PL involved in signal transmission during cell growth and differentiation.

PGs are present in high concentration in mitochondrial membranes and as components of pulmonary surfactant. PG is synthesized from CDP-diacylglycerol and glycerol-3-phosphate. The critical role of PG is to serve as the precursor for the synthesis of DPGs or cardiolipin. DPGs are synthesized by the condensation of CDP-diacylglycerol with PG, and are also found primarily in the mitochondrial membrane.

The backbone of SM is sphingosine, an amino alcohol formed from palmitate and serine. The amino terminal is acylated with a long-chain acyl coenzyme A (CoA) to give ceramide. Further esterification of the end hydroxyl group by phosphocholine forms SM.

Little is known currently about the biosynthesis or function of the unusual phosphonolipids. Fatty acid composition data indicate that animal and plant tissues synthesize phosphonolipids from an entirely different fatty acid pool to other common membrane lipids (Mukhamedova & Glushenkova, 2000).

B. Biological functions of phospholipids

PLs are important in providing the permeable barrier of the cell and intracellular organelles, in forming a membrane surface for biochemical reactions catalysed by membrane bound proteins, in signal transduction in response to external or internal stimuli, and they serve as precursors for the synthesis of other bioactive molecules. Although much has been learned, there still is a great need to conduct further basic research to understand the biological functions of PLs fully.

1. Membrane lipids

The most important biological function of PLs is that they form membranes and provide cellular integrity. PLs are quantitatively the most important lipids in all cell and organelle membranes, with glycosphingolipids and cholesterol of secondary importance. Sphingolipids are important components of some cellular membranes, such as in the central nerve system. Membrane lipids are amphiphilic, with a hydrophilic head group and lipophilic fatty acyl tails. They spontaneously form an ordered bilayer structure when suspended in an aqueous environment. The lipophilic tails line up facing each another and hydrophilic heads face both sides of the membrane (see Figures 5.18 and 5.19). This membrane is partially permeable and semi-fluid, and embedded proteins and phospholipid molecules are able to move within it to a certain degree. Such structure and movement can be described by the Fluid Mosaic Model that describes the membrane as a mosaic of lipid molecules that act as a 'solvent' for other substances and proteins, so proteins and lipid molecules are free to diffuse laterally through the matrix. Therefore, biological membrane can be thought of as a lipid–protein matrix in which specific biochemical and biological functions are performed by the peripheral or integral proteins as enzymes and transporters, and the lipids provide a permeable barrier and structural integrity.

Biological membranes are diverse, and contain various ratios of lipid (25–79%), protein (18–75%), and carbohydrate (0–8%), depending on their source (Vance *et al.*, 1993). The common plasma membranes enclose the cytoplasm of all cells and also separate internal compartments or organelles in which essential biochemical reactions occur. The eukaryotic organelles have specific functions, such as lysosomes for digestion, mitochondria for respiration, chloroplasts for photosynthesis, and endoplasmic reticulum and Golgi apparatus for secretion. Thus, they all have their specialized membrane systems and lipid composition as shown in Table 1.1.

The physical properties of biological membranes are mostly affected by temperature, lipid composition and extra-membrane interactions. Phase transition temperature, the temperature at which the membrane changes from the highly ordered gel phase to the random liquid crystalline phase, is lowered by short-chain fatty acid and unsaturation. The nature of the head group also affects transition temperature; for example, PA and PE dipalmitoyl esters have higher transition temperatures than PC and PG. Divalent ions (Ca^{2+} and Mg^{2+}),

Table 1.1. Lipid composition (%) of rat membrane preparations[a]

Source	Chol-esterol	PC	SM	PE	PI	PS	PG	DPG	PA	Glyco-lipids	Others
Liver											
Cytoplastic	30	18	14	11	4	9			1		13
Mitochondria (inner)	<3	45	2	25	6	1	2	18			
Mitochondria (outer)	<5	50	5	23	13	2	2	3	1		
Nuclear	10	55	3	20	7	3			1		1
Golgi	8	40	10	15	6	4					17
Lysosomes	14	25	24	13	7	7		5			5
Brain											
Myelin	22	11	6	14		7				21	19
Synaptosome	20	24	4	20	2	8			1		21
Erythrocyte	24	31	9	15	2	7				3	9

[a]From Vance *et al.*, 1993. Abbreviations: PC = phosphatidylcholine; SM = sphingomyelin; PE = phosphatidylethanolamine; PI = phosphatidylinositol; PS = phosphatidylserine; PG = phosphatidylglycerol; DPG = diphosphatidylglycerol (cardiolipin); PA = phosphatidic acid.

by ionic interaction with phosphate, tend to bind PLs molecules together, and decrease molecular mobility and membrane fluidity, increase transition temperature and thus stabilize the membrane (MacDonald *et al.*, 1976; Asai & Watanabe, 1999). A more detailed discussion of these physical properties is to be found in Chapter 5.

2. *PLs provide polyunsaturated fatty acids for eicosanoid biosynthesis*

The fatty acid distribution at the *sn*-1 and *sn*-2 positions of PLs is continually changing, because of the PL degradation and the continuous PL modification by phospholipases. Phospholipase A_1 and phospholipase A_2 are responsible for these changes. Phospholipase A_2 is a particularly important enzyme, because it catalyses the release of the highly unsaturated long-chain fatty acid on the *sn*-2 position, and these fatty acids are then used as substrates for eicosanoid synthesis. The *sn*-2 fatty acid composition in comparison to *sn*-1 has been shown by many researchers to be more unsaturated. Studies by Wang and co-workers (1997) and Wang and Hammond (1999) on soybean PLs offer examples in which the unsaturated fatty acids are mainly located on the *sn*-2 position of the PL molecules, while the saturated fatty acids are more concentrated on the *sn*-1 position.

Eicosanoids are a class of oxygenated fatty acids that act as local hormones which can be released by most cells. The synthesis of eicosanoids is triggered by cellular stimuli, and the free 20-carbon essential fatty acid released by a phospholipase is a precursor for the synthesis of four families of eicosanoids: the prostaglandins, the prostacyclins, the thromboxanes and the leukotrienes. For each, there are two or three separate series, derived either from a ω-3 or ω-6 essential fatty acid. The different activities from these various acids largely explain the different health effects of ω-3 and ω-6 fatty acids. The ω-6 eicosanoids are generally pro-inflammatory but ω-3 eicosanoids typically have an opposite effect. These hormone-like compounds play important roles in inflammation, immunity, and as messengers in the central nervous system.

When activated, mast cells generate prostaglandins and leukotrienes from cell membrane arachidonic acid by cyclooxygenase isoenzymes and 5-lipoxygenase, and these diverse lipid mediators can initiate, amplify or dampen inflammatory responses and influence the magnitude, duration and nature of subsequent immune responses (Boyce, 2007). The prime mode of prostaglandin and leukotriene action is through specific G protein-coupled receptors, many of which have been cloned recently. Important insights into the mechanisms of inflammatory responses, pain and fever have been obtained from current understanding of eicosanoid biology (Funk, 2001). The biological functions of all eicosanoids are among the most complex in the human body.

3. PC as source of choline and its role in homocysteine metabolism and vascular disease

Lecithin or PC are a good source of choline, an essential nutrient. Choline is a precursor in the synthesis of the neurotransmitter, acetylcholine. Choline prevents fat accumulation in the liver, and choline deficiency disturbs lecithin synthesis which is needed to export triacylglycerols from the liver as part of lipoproteins (Orthoefer and List, 2006). Lecithin lowers serum cholesterol levels. The hypocholesterolemic and antiatherogenic effects of soybean lecithin (3.4% in diet) were shown in monkeys and hamsters, while the HDL (high density lipoprotein)-cholesterol was not altered (Wilson *et al.*, 1998).

Healthy humans fed diets deficient in choline develop liver dysfunction that is similar to that seen in choline-deficient animals. Consuming a choline-deficient diet for 3 weeks, humans had depleted stores of choline in tissues and had developed signs of incipient liver dysfunction (Zeisel *et al.*, 1991). The observations support the conclusion that choline is an essential nutrient for humans when excess methionine and folate are not available in the diet. In experimental animals, fatty liver occurs through choline deficiency because PC synthesis is required for removing VLDL (very low density lipoprotein). This accumulation of lipids in liver may explain why choline-deficient rats spontaneously develop hepatocarcinoma. Choline deficiency has also been associated with the accumulation of 1,2-diacylglycerol, an activator of protein kinase C. Several lines of evidence indicate that cancers might develop, in addition to the abnormalities in protein kinase C-mediated signal transduction (Zeisel, 1992).

Choline is also a major source of the methyl groups involved in the formation of methionine from homocysteine, a sulfur-containing amino acid ($HSCH_2CH_2CH[NH_2]COOH$) that is implicated in cardiovascular disease risk. High homocysteine levels increase the risk of cardiovascular disease, and medical therapy with choline can reduce homocysteine levels (Innis *et al.*, 2007; Da Costa *et al.*, 2005; Zeisel, 2005). The atherogenic effect of homocysteine was discovered by analysis of the pathology of genetic disorders of homocysteine metabolism; high homocysteine levels caused arteriosclerotic plaques in experimental animals, and epidemiological studies have also shown that hyperhomocysteinemia is a major risk factor for human arteriosclerosis (McCully, 1998). Betaine, formed by choline oxidation, transfers a methyl group to homocysteine to form methionine by a transmethylation pathway in liver.

4. Choline in memory and brain function

It is widely recognized that with ageing, spatial learning and memory ability decline in both rats and humans. When young rats were given perinatal choline supplementation, the memory function improvement appeared to be permanent, as shown by their good memory when the same rats were in old age (Meck, 1998). Other animal and human studies have also shown that lecithin improves memory and learning. When aged male rats were given lecithin, they showed markedly higher spatial memory than untreated rats as demonstrated in a water maze. The acetylcholine content of brains of such rats was higher than those of controls. Studies of behavioural function involving short-term memory and learning in mice showed that the negative effect of a high ω-6 fatty acid diet can be overcome by replacing the linoleic acid with arachidonic acid (Wainwright *et al.*, 1997).

Lecithin and choline are essential in brain and mental development in the foetus and infant. Choline supplements administered to rats *in utero* or shortly after birth permanently alter brain function (Zeisel, 1992). Choline in mother's milk is at a level of 100 times above that in the maternal bloodstream, and many vital organs, such as the central nerve system, kidney and liver, contain high levels of PLs (Orthoefer & List, 2006).

5. *PL oxidation related to ageing and as antioxidant*

Major membrane changes in ageing include increases in membrane rigidity and levels of cholesterol, PC, ω-6 PUFA, and decreases in content of ω-3 PUFA, mitochondrial cardiolipin and antioxidants such as coenzyme Q10 (Pepe *et al.*, 2002). These alterations have a significant effect on the functions of membrane proteins involved in signal transduction, redox reactions and oxidative phosphorylation. Fortunately, some of the age-related changes can be beneficially modified by dietary strategy. Dietary restriction significantly counteracted most of the age-related PL changes occurring in the rat liver, kidney and heart (Tamburini *et al.*, 2004). A diet rich in ω-3 LC-PUFA can reverse the membrane PL ω-6: ω-3 PUFA imbalance, thus, dietary modification of PLs may have a significant impact on health (Wainwright *et al.*, 1997).

Alzheimer's disease is associated with oxidative stress, which can be detected *in vivo* by measuring plasma lipoperoxides and TBARS levels and the degree of glutathione oxidation in red blood cells (Vina *et al.*, 2004). The lipid oxidation was attributed to the reaction of the unsaturated fatty acids in membrane phospholipids. Antioxidant vitamin E was shown to improve cognitive performance in some patients (Vina *et al.*, 2004). The effects of age-related oxidative damage on PUFA in membrane PLs, and α-tocopherol concentration were also investigated in mice, and such oxidation was again related to Alzheimer's disease (Petursdottir *et al.*, 2007).

Oxidation-resistant membrane PLs can explain longevity differences among the longest-living rodents and similarly-sized mice (Hulbert *et al.*, 2006). Normal mice had 9 times more docosahexaenoic acid (DHA) in cell membranes than the naked mole-rat membranes. The naked mole-rat had a longer lifespan, suggesting that membrane PL composition is an important determinant of longevity.

Oxidative stress was proved to be involved in the pathogenesis of age-related macular degeneration because of the oxidation of PLs in the photoreceptors, similar to the PL oxidation in the arterial intima of patients with atherosclerosis (Suzuki *et al.*, 2007). Therefore, controlling oxidation of PLs may be a potential treatment for age-related macular degeneration.

Oxidized PLs also showed a strong association with the presence and extent of coronary artery disease, and the combined effect of hypercholesterolaemia and the oxidized PLs significantly increased the probability of coronary artery disease among patients 60 years of age or younger. Such atherogenicity may be mediated in part by the pro-inflammatory oxidized PLs (Tsimikas *et al.*, 2005).

Mixtures of PLs in a commercial form (not highly pure) is referred to as lecithin. Lecithin is an effective antioxidant *in vitro*. The antioxidant property of crude soybean lecithin has been studied in various storage tests with sunflower oil and lard (Nasner, 1985). The addition of a very small quantity of phospholipids after oil refining improved oxidative stability, and PLs at 100 ppm in salmon oil had a stabilizing effect (King *et al.*, 1992). The antioxidant activity depends on the composition of the PLs, and the tocopherol content of the oil. The synergistic effects of tocopherol, ascorbate and lecithin have been clearly shown in fish oil oxidation (Hamilton *et al.*, 1998; Segawa *et al.*, 1995).

6. *Role of cardiolipin in mitochondrial physiology and disease*

Cardiolipin is found almost exclusively in mitochondria membrane and in bacteria. It can account for as much as 20% of mitochondrial lipids. Interestingly, mammalian cardiolipin contains up to 90 mol% of one fatty acid, linoleic acid, which gives a high degree of structural uniformity. Mitochondrial membrane has a very ordered structure and contains a distinct number of cardiolipin molecules firmly imbedded in the membrane (Schlame *et al.*, 2000). Researchers have tried to isolate mutants lacking the 'essential' cardiolipin to study physiological consequence to a eukaryotic cell when cardiolipin is missing. These mutants can elucidate the role of cardiolipin in oxidative phosphorylation and also provide insights of other potential functions of this lipid.

Barth Syndrome is a rare genetic disorder that has a mutation in the gene coding for an enzyme involved in the biosynthesis of cardiolipin. This results in abnormal mitochondria and poor production of ATP. Cardiomyopathy and general weakness is common to these patients. A comprehensive review on the function and regulation of the biosynthesis and resynthesis of mitochondrial PLs in the mammalian heart demonstrates the relationship between mitochondrial PL composition and Barth Syndrome (Hatch, 2004). In mitochondria-rich heart, an alteration in the content and fatty acid composition of the heart PLs was linked to alterations in myocardial electrical activity (Hatch, 2004). Cardiolipin may be involved in basic mechanism of energy transduction (Dowhan, 1997).

Cardiolipin may also be involved in ageing for which a decline of mitochondrial function is one of the early events. Age-related loss of cardiolipin is often shown in mitochondria from rat heart and liver, and in epidermal cells from humans. This is associated with decreased activities of mitochondrial enzymes. Cardiolipin oxidation and its loss in rat brain mitochondria was associated with impaired activity of the electron transport chain, leading to neuronal damage associated with brain ageing and neurodegenerative disorders (Sen *et al.*, 2006). Cardiolipin mainly has polyunsaturated acyl groups susceptible to oxidative damage. Therefore, the age-related accumulation of free radicals may be responsible for the decline of cardiolipin due to oxidation.

Recently, a new arena in cardiolipin oxidation has emerged. Mitochondria and their role as the major intracellular source of reactive oxygen species were linked to regulation of cell death by oxidation of macromolecules that has been implicated in mtDNA mutations and ageing (Ott *et al.*, 2007). These reactive oxygen species play an important role in the release of cytochrome c and other pro-apoptotic proteins, which can trigger caspase activation and apoptosis. Oxidation of cardiolipin reduces cytochrome c binding and results in an increased level of 'free' cytochrome c in cells.

7. Membrane lipid uneven distribution and cell function

The ubiquitous phenomenon of membrane PL asymmetry (i.e., the unequally distributed PLs between the inner and outer leaflets of many biological membranes, such as the mitochondria membranes as shown in Table 1.1) is characterized by a different lipid composition between the inner and outer leaflet of the membrane bilayer (Bevers *et al.*, 1998). Transbilayer asymmetry is controlled by different membrane proteins that function as lipid transporters, catalysing uni- or bi-directional transbilayer movement of lipids. Under normal conditions, an ATP-dependent protein generates and maintains phospholipid asymmetry by promoting unidirectional transport of amino-PLs from the outer to the inner leaflet. The membrane lipid asymmetry may be compromised during cellular activation by a Ca^{2+}-dependent transporter that facilitates rapid bi-directional movement of all major phospholipid classes (Bevers *et al.*, 1998). A major consequence of this collapse of lipid asymmetry is the exposure of PS at the outer membrane surface. Surface exposure of PS has important physiological and pathological implications for blood coagulation, apoptosis and cell–cell recognition (Bevers *et al.*, 1998). The generation and maintenance of red blood cell membrane PL asymmetry and the involvement of flippase are discussed by Roelofsen and co-workers (1996).

In membranes such as the endoplasmic reticulum, PLs can move readily across the bilayer, assisted by membrane proteins that facilitate a passive equilibration of lipids. In the plasma membrane, however, distribution of PLs is regulated by lipid flippases, and this redistribution has important functions, such as in release of lipids from cells, and in vesicular traffic in the process of vesicle formation (Pomorski *et al.*, 2004).

8. Anionic PLs in blood coagulation and thrombosis

It is known that physiological blood coagulation requires the presence of membranes composed of negatively charged PLs. Zymogen activations occur rapidly when the vitamin K-dependent enzyme binds to a cofactor to activate the protein substrate. In addition to a net negative charge, the nature of the PL head group appears to contribute to catalytic and binding efficiency. PS is generally considered to be the most important PL in blood clotting, and the majority of biophysical and kinetic studies of the assembly of the vitamin K-dependent complexes have used membranes composed solely of PC and PS.

Anionic PLs, mainly PS present in the cytoplasmic leaflet of the plasma membrane, play a critical role not only in blood coagulation, but also in thrombosis (Bevers *et al.*, 2003). Platelets form physical barriers upon aggregation in order to prevent further blood loss, and also contribute to the propagation of the blood coagulation process. The surface activity of platelet coagulant is due to the anionic PLs (Bevers *et al.*, 2003). It has been recently found that the presence of antiphospholipid autoantibodies (aPL) may play a role in the antiphospholipid syndrome which is characterized by thrombosis or pregnancy morbidity (Lopez-Pedrera *et al.*, 2006). The mechanisms by which aPL cause thrombosis are not completely understood. There is no unique mechanism to explain the presence of aPL, though aPL are able to recognize, injure and to bind anionic PLs which are exposed when platelets are activated (Lopez-Pedrera *et al.*, 2006). Blood clotting is also affected by membrane-bound enzymes and it is known that membrane environment or composition can affect enzymatic activity. It was shown that full proteolytic activity of the ectoenzyme that triggers blood clotting required an extremely high local concentration of anionic PLs (Shaw *et al.*, 2007).

9. *Sphingomyelin and its cellular function*

SM belongs to both the sphingolipid and PL classes, and it is present in all eukaryotic cell membranes. It is particularly concentrated in the nervous system, and is the major component of myelin, the fatty material wrapped around nerve cells. Certain diseases lead to the deterioration of the myelin sheath, resulting in impaired function of the nervous system.

Sphingolipids are generally present in much lower concentration than PLs in membranes. They and their metabolites (ceramide, sphingosine, sphingosine-1-phosphate and sphingosylphosphorylcholine) are all highly bioactive. They have been found to act as mediators of cell growth, cell differentiation and programmed cell death (apoptosis). *In vitro* studies showed that ceramide and sphingosine were toxic for a variety of transformed cell lines and even inhibit cell transformation during the early events of carcinogenesis (Merrill & Schmelz, 2001).

Sphingolipids reduce the risk of colon cancer as shown by Dillehay and co-workers (1994). The occurrence of aberrant colonic crypt foci (ACF, the early biomarker of tumour development) of mice fed sphingomyelin at 0.05% of their diets was reduced by 50% in comparison to the control group. In a longer term study, Schmelz and co-workers (1996) showed ACF was reduced in mice fed sphingomyelin (0.1% of diet) by up to 70%. Although the incidence of colonic tumours was not reduced in the sphingomyelin-fed mice, the proportion of benign adenomas versus adenocarcinomas was higher in these mice than in the control group. It was suggested that sphingomyelin may prevent adenomas from progressing into adenocarcinomas. Sphingolipids, such as glucosylceramide and lactosylceramide, also were shown to reduce ACF by 50–80%, indicating that sphingolipids suppress colon carcinogenesis through the release of their metabolites by hydrolysis (Schmelz, 2000).

Sphingolipids reduced hepatic cholesterol content in a short-term rat feeding study (Imaizumi *et al.*, 1997) by decreasing cholesterol absorption and/or increasing faecal excretion. In a long-term feeding study of two generations, rats fed sphingolipids at 1% of their diet have their total plasma cholesterol reduced by 30% (Kobayashi *et al.*, 1997). Dietary sphingolipids may influence plasma and liver lipid levels in humans, but more research is needed to understand the mechanism of cholesterol reduction better.

Sphingolipids offer protection against pathogenic microorganisms (bacteria and viruses) and toxins. Synthetic sphingolipids have been successfully used to prevent bacterial and viral infections by binding to pathogens and removing them from the intestine (Vesper *et al.*, 1999). The primary compound in human milk that protects against pathogens is assumed to be glycosphingolipids (Vesper *et al.*, 1999). Sphingolipids are also found to reduce skin carcinoma development (Birt *et al.*, 1998).

10. PLs in nuclear events

PLs also play a role in nuclear events (Alessenko & Burlakova, 2002). Nuclei have a very active lipid metabolism, which seems to play an important role in the transduction of the signals to the genome in response to agonists at the plasma membrane level. There is evidence of the presence of PL-calcium-dependent protein kinase C (PKC) in nuclei interacting with nuclear phosphoinositol and SM products (Alessenko & Burlakova, 2002).

11. PLs as a source of second messengers for cell function

Lipid second messengers derived from glycerolipids and sphingolipids play an important role in smooth muscle function (Ohanian *et al.*, 1998). The PLs located within the plasma membrane act as a reservoir for signalling molecules. When a receptor binds its appropriate activating ligand, a chain of events is initiated which leads to the breakdown of these lipids by phospholipases and the release of second messengers. Ohanian and co-workers (1998) examined the profile of lipid second messengers derived from glycerophospholipids (PI and PC) and SM. Ceramide and sphingoid base were formed in response to vasoconstrictor hormones, and they played a role in vascular smooth muscle cells.

PA can act as an intracellular mediator and extracellular messenger for biological activities once hydrolysed by enzymes (English *et al.*, 1996), and such activities can be attributed to lyso-PA and diacylglycerol generated by cellular hydrolases. Diacylglycerol has been recognized as a potent second messenger and can also be generated via the cleavage of inositol- or choline-containing PLs, and it was involved in the transduction of proliferative signals (Wickremasinghe *et al.*, 1993). However, in the regulation of leukaemia cell proliferation, inositol lipids rather than choline lipids were the second messenger molecules with biological functions (Wickremasinghe *et al.*, 1993). Natural killer cells kill tumour target cells by exocytosis of cytotoxic granules via a calcium dependent process, in which inositol trisphosphate and diacylglycerol functioned as second messengers (Atkinson *et al.*, 1989). The study of PL metabolism and the related second messenger systems in the selective neuronal damage after ischaemia in gerbils also showed that inositol PL was the second messenger (Abe *et al.*, 1992). PI monophosphate and bisphosphate have been shown to be key intermediates in stimulus response pathways in many animal cells, and these lipids and related enzymes were also found in the plasma membranes of plant cells (Chen *et al.*, 1991). The study of PL signalling in responses to environmental stresses such as abscisic acid (ABA) exposure showed that ABA evoked 1.7-fold increase in PA and sevenfold increase in diacylglycerol pyrophosphate (DGPP); therefore, DGPP was regarded as the PL second messenger in ABA signalling (Zalejski *et al.*, 2005).

A comprehensive review of PL-derived second messengers and their roles in the control of cellular function and in cell communication describes the modification of inositol PL, PC and SM and their functional significance (Exton, 1998).

12. PLs in pulmonary function

PLs are involved in physiological functions in the cell membranes, and more importantly as constituents of pulmonary surfactant. Dipalmitoyl PC (DPPC) is the major component of pulmonary surfactant, and is responsible for the high surface pressure, although other saturated and unsaturated PLs as well as surfactant proteins also play a role in the dynamic surface activity of the pulmonary film (Hsu, 1997). Exogenous surfactant is a specialized biomaterial used for substitution of the lipoprotein mixture normally present in lungs (pulmonary surfactant). This is used in the treatment of Respiratory Distress Syndrome, a disease of preterm infants mainly caused by a deficiency of mature lung surfactant (Banerjee *et al.*, 2000). Pulmonary surfactant is known to stabilize small alveoli and prevent them from collapsing during expiration due to its unique surface properties. The composition of surfactant composition affects pulmonary functions. The PL composition and biophysical behaviour of pulmonary surfactant change in different types of calves. Higher levels of DPPC seemed to relate to higher resistance to respiratory troubles (Danlois *et al.*, 2003).

13. Phosphonolipids and their possible function

Phosphonolipids are found in fish, certain mammals, various bacteria strains, egg yolk (1% phosphonolipids in total PLs, 98.1% of which is phosphono-PE and 1.9%, phosphono-PC) and insects (Mukhamedova & Glushenkova, 2000). Marine invertebrates are probably the richest sources of these lipids (Itonori *et al.*, 2007). Phosphonolipids from plant sources are much less studied, possibly because of their low concentration and unknown function. The unique fatty acid composition of phosphonolipids may indicate that they have a unique role in membranes, even though the biological function of these lipids is not clearly understood (Mukhamedova & Glushenkova, 2000). Many researchers think that phosphonolipids, like PLs, are cell-membrane structural elements, and they also participate in membrane stabilization. Phosphonolipids are resistant to hydrolytic degradation by enzymes such as phospholipase. The presence of a P–C bond in phosphonolipids inactivates phospholipase D and certain phosphonolipids inhibit phospholipase C and other hydrolytic enzymes (Mukhamedova & Glushenkova, 2000).

Recently, there have been some interesting applications using synthetic cationic phosphonolipids as carriers in a synthetic transfection system for DNA transfer and gene therapy (Koumbi *et al.*, 2006; Montier *et al.*, 2003; Floch *et al.*, 2000).

References

Abe, K, Araki, T, Kawagoe, J, Aoki, M and Kogure, K (1992) Phospholipid metabolism and second messenger system after brain ischemia. *Advances in Experimental Medicine and Biology*, 318, 183–95.

Alessenko, AV and Burlakova, EB (2002) Functional role of phospholipids in the nuclear events. *Bioelectrochemistry*, 58, 13–21.

Asai, Y and Watanabe, S (1999) Effect of divalent cations on the membrane properties of the lipid A analog E5531. *Drug Development and Industrial Pharmacy*, 25, 1107–1113.

Atkinson, EA, Gerrard, JM, Hildes, GE and Greenberg, AH (1989) Production of inositol-phospholipid-derived second messengers in a rat natural killer cell line exposed to susceptible tumor targets. *Natural Immunity and Cell Growth Regulation*, 8, 223–30.

Banerjee, R, Puniyani, RR and Bellare, JR (2000) Analysis of dynamic surface properties of therapeutic surfactants and lung phospholipids. *Journal of Biomaterials Applications*, 15, 140–159.

Bevers, EM, Comfurius, P, Dekkers, DWC, Harmsma, M and Zwaal, RFA (1998) Transmembrane phospholipid distribution in blood cells. Control mechanisms and pathophysiological significance. *Biological Chemistry*, 379, 973–986.

Bevers, EM, Wolfs, JL, Comfurius, P and Zwaal, RFA (2003) Platelets and propagation of coagulation: How platelets control surface-exposed procoagulant phospholipids. In: *Thrombosis: Fundamental and Clinical Aspects* (J Arnout, ed), ETRO Advanced Teaching Course on Thrombosis, Blankenberge, Belgium, Oct. 25–31, pp. 143–159.

Birt, DA, Merrill, AH, Barnett, T. Enkvetchakul, B, Pour, PM, Liotta, DC, Geisler, V, Menaldino, DS and Schwartzbauer, J (1998) Inhibition of skin carcinomas but not papillomas by sphingosine, N-methylsphingosine, and N-acetylsphingosine. *Nutr. Cancer*, 31, 119–126.

Boyce, JA (2007) Mast cells and eicosanoid mediators: a system of reciprocal paracrine and autocrine regulation. *Immunological Reviews*, 217, 168–185.

Chen, Q, Brglez, I and Boss, WF (1991) Inositol phospholipids as plant second messengers. *Symposia of the Society for Experimental Biology*, 45, 159–75.

Da Costa, K, Gaffney, CE, Fischer, LM and Zeisel, SH (2005) Choline deficiency in mice and humans is associated with increased plasma homocysteine concentration after a methionine load. *American Journal of Clinical Nutrition*, 81, 440–444.

Danlois, F, Zaltash, S, Johansson, J, Robertson, B, Haagsman, HP, Rollin, F, Ruysschaert, JM and Vandenbussche, G (2003) Pulmonary surfactant from healthy Belgian white and blue and Holstein Friesian calves: Biochemical and biophysical comparison. *Veterinary Journal*, 165, 65–72.

Dillehay, DL, Webb, SK, Schmelz, EM and Merrill, AH (1994) Dietary sphingomyelin inhibits 1,2-dimethylhydrazine-induced colon cancer in CF1 mice. *J. Nutr.*, 124, 614–620.

Dowhan, W (1997) The role of phospholipids in cell function. In: *Advances in Lipobiology* (RW Gross, ed), Jai Press Inc., Greenwich, Connecticut, pp. 79–108.

English, D, Cui, Y and Siddiqui, RA (1996) Messenger functions of phosphatidic acid. *Chemistry and Physics of Lipids*, 80, 117–132.

Exton, JH (1998) Phospholipid-derived second messengers. In: *Handbook of Physiology, Section 7: The Endocrine System*, Vol 1 (PM Conn, ed), American Physiology Society, Oxford University Press Inc, NY, 255–291.

Floch, V, Loisel, S, Guenin, E, Herve, AC, Clement, J, Yaouanc, J, des Abbayes, H and Ferec, C (2000) Cation substitution in cationic phosphonolipids : A new concept to improve transfection activity and decrease cellular toxicity. *Journal of Medicinal Chemistry*, 43, 4617–4628.

Funk, CD (2001) Prostaglandins and leukotrienes: Advances in eicosanoid biology. *Science*, 294, 1871–1875.

Hamilton, RJ, Kalu, C, McNeil, GP, Padley, FB and Pierce, JH (1998) Effects of tocopherols, ascorbyl palmitate, and lecithin on autoxidation of fish oil. *J. Am. Oil Chem. Soc.*, 75, 813–822.

Hatch, GM (2004) Cell biology of cardiac mitochondrial phospholipids. *Biochemistry and Cell Biology*, 82, 99–112.

Hsu, S (1997) Surface properties of phospholipids: Pulmonary surfactant perspective. *Jiemian Kexue Huizhi*, 20, 1–11.

Hulbert, A J, Faulks, SC and Buffenstein, R (2006) Oxidation-resistant membrane phospholipids can explain longevity differences among the longest-living rodents and similarly-sized mice. *Journals of Gerontology, Series A: Biological Sciences and Medical Sciences*, 61A, 1009–1018.

Imaizumi, K, Tominaga, A, Sato, M and Sugano, M (1997) Effects of dietary sphingolipids on levels of serum and liver lipids in rats. *Nutr. Res.*, 12, 111–114.

Innis, SM. Davidson, A, George, F, Melynk, S and James, SJ (2007) Choline-related supplements improve abnormal plasma methionine-homocysteine metabolites and glutathione status in children with cystic fibrosis. *American Journal of Clinical Nutrition*, 85, 702–708.

Itonori, S, Kitamura, T, Tanaka, R, Miyagaki, N, Saito, H and Sugita, M (2007) Aliphatic components of phosphonolipids (ceramide 2-aminoethylphosphonates) from the pearl oyster, *Pinctada martensii. Shiga Daigaku Kyoikugakubu Kiyo, III: Shizen Kagaku*, 56, 51–62.

King, MF, Boyd, LC and Sheldon, BW (1992) Antioxidant properties of individual phospholipids in a salmon oil model system. *J. Am. Oil Chem. Soc.*, 69, 545–551.

Kobayashi, T, Shimizugawa, T, Osakabe, T, Watanabe, S and Okuyama H (1997) A long-term feeding of sphingolipids affected the levels of plasma cholesterol and hepatic triacylglycerol but not tissue phospholipids and sphingolipids. *Nutr. Res.*, 17, 111–114.

Koumbi, D, Clement, J, Sideratou, Z, Yaouanc, J, Loukopoulos, D and Kollia, P (2006) Factors mediating lipofection potency of a series of cationic phosphonolipids in human cell lines. *Biochimica et Biophysica Acta, General Subjects*, 1760, 1151–1159.

Lopez-Pedrera, C, Buendia, P, Barbarroja, N, Siendones, E, Velasco, F and Cuadrado, MJ (2006) Antiphospholipid-mediated thrombosis: Interplay between anticardiolipin antibodies and vascular cells. *Clinical and Applied Thrombosis/Hemostasis*, 12, 41–45.

MacDonald, RC, Simon, SA and Baer, E (1976) Ionic influences on the phase transition of dipalmitoylphosphatidylserine. *Biochemistry*, 15, 885–891.

McCully, KS (1998) Homocysteine and vascular disease: The role of folate, choline, and lipoproteins in homocysteine metabolism. In: *Choline, Phospholipids, Health, and Disease* (SH Zeisel and BF Szuhaj, eds), 7th International Congress on Phospholipids, Brussels, AOCS Press, Champaign, IL, pp. 117–130.

Meck, WH (1998) Choline and development of brain memory functions across the life span. In: *Choline, Phospholipids, Health, and Disease* (SH Zeisel and BF Szuhaj, eds), 7th International Congress on Phospholipids, Brussels, AOCS Press, Champaign, IL, p. xi.

Merrill, AH and Schmelz, EM (2001) Sphingolipids: mechanism-based inhibitors of carcinogenesis produced by animals, plants, and other organisms. In: *Handbook of Nutraceuticals and Functional Foods* (RC Wildman, ed), CRC Press, New York, NY, pp. 377–392.

Montier, T, Delepine, P, Le Ny, K, Blanc, F, Fichou, Y, Le Bris, M, Gillet, D, Piquet, E, Clement, JC, Yaouanc, JJ, des Abbayes, H and Ferec, C (2003) Cationic phosphonolipids' evolution: a constant and continuous improvement of transfection activity. *Recent Research Developments in Chemistry*, 1, 41–58.

Mukhamedova, KS and Glushenkova, AI (2000) Natural phosphonolipids. *Chem. Nat. Compounds*, 36, 329–341.

Nasner, A (1985) Antioxidative properties of lecithin. *Fette, Seifen, Anstrichmittel*, 87, 477–81.

Ohanian, J, Liu, G, Ohanian, V and Heagerty, AM (1998) Lipid second messengers derived from glycerolipids and sphingolipids, and their role in smooth muscle function. *Acta Physiologica Scandinavica*, 164, 533–548.

Orthoefer, FT and List, GR (2006) Phospholipids/lecithin: A class of neutraceutical lipids. In: *Nutraceutical Science and Technology, Vol 5: Nutraceutical and Specialty Lipids and their Co-Products* (F Shahidi, ed), Taylor and Francis, St. John's, Newfoundland, pp. 509–530.

Ott, M, Gogvadze, V, Orrenius, S and Zhivotovsky, B (2007) Mitochondria, oxidative stress and cell death. *Apoptosis*, 12, 913–922.

Pepe, S, Rosenfeldt, FL, Lakatta, EG and Sollott, SJ (2002) Mitochondrial membrane alterations by ageing and diet: impact on recovery of myocardial function after stress. *Adaptation Biology and Medicine*, 3, 83–92.

Petursdottir, AL, Farr, SA, Morley, JE, Banks, WA and Skuladottir, GV (2007) Lipid peroxidation in brain during ageing in the senescence-accelerated mouse (SAM). *Neurobiology of Aging*, 28, 1170–1178.

Pomorski, T, Holthuis, J CM, Herrmann, A and Van Meer, G (2004) Tracking down lipid flippases and their biological functions. *Journal of Cell Science*, 117, 805–813.

Roelofsen, B, Middelkoop, E, Vermeulen, WP, Smith, AJ and Op den Kamp, JAF (1996) Phospholipid flippases: neither exclusively, nor only involved in maintaining membrane phospholipid asymmetry. *Cell Biology*, 96, 367–382.

Schlame, M, Rua, D and Greenberg, ML (2000) The biosynthesis and functional role of cardiolipin. *Prog Lipid Res.*, 39, 257–88.

Schmelz, EM (2000) Dietary sphingomyelin and other sphingolipids in health and disease, *Nutr. Bull.*, *25*, 135–139.

Schmelz, EM, Dillehay, DL, Webb, SK, Reiter, A, Adams, J and Merrill, AH (1996) Sphingomyelin consumption suppresses aberrant colonic crypt foci and increases the proportion of adenomas versus adenocarcinomas in CF1 mice treated with 1,2-dimethylhydrazine: implications for dietary sphingolipids and colon carcinogenesis. *Cancer Res.*, 56, 4936–4941.

Segawa, T, Kamata, M, Hara, S and Totani, Y (1995) Antioxidant activity of phospholipids for polyunsaturated fatty acids of fish oil. III. Synergism of nitrogen-containing phospholipids with tocopherol. *Yukagaku*, 44, 36–42.

Sen, T, Sen, N, Tripathi, G, Chatterjee, U and Chakrabarti, S (2006) Lipid peroxidation associated cardiolipin loss and membrane depolarization in rat brain mitochondria. *Neurochemistry International*, 49, 20–27.

Shaw, AW, Pureza, VS, Sligar, SG and Morrissey, JH (2007) The local phospholipid environment modulates the activation of blood clotting. *Journal of Biological Chemistry*, 282, 6556–6563.

Suzuki, M, Kamei, M, Itabe, H, Yoneda, K, Bando, H and Kume, N, Tano, Y (2007) Oxidized phospholipids in the macula increase with age and in eyes with age-related macular degeneration. *Molecular Vision*, 13, 772–778.

Tamburini, I, Quartacci, MF, Izzo, R and Bergamini, E (2004) Effects of dietary restriction on age-related changes in the phospholipid fatty acid composition of various rat tissues. *Aging: Clinical and Experimental Research*, 16, 425–431.

Tsimikas, S, Brilakis, ES, Miller, ER, McConnell, JP, Lennon, RJ, Kornman, KS, Witztum,

JL and Berger, PB (2005) Oxidized phospholipids, Lp(a) lipoprotein and coronary artery disease. *New England Journal of Medicine*, 353, 46–57.

Vance, DE, Jacobson, GR and Saier, Jr. MH (1993) Lipids and membranes. In: *Biochemistry* (G Zubay, ed), 3rd edn, WCB Publishers, Dubuque, IA, pp. 163–196.

Vesper, H, Schmelz, EM, Nikolova–Karakashian, MN, Dillehay, DL, Lynch, DV and Merrill, AH (1999) Sphingolipids in food and the emerging importance of sphingolipids to nutrition. *J. Nutr.*, 129, 1239–1250.

Vina, J, Lloret, A, Orti, R and Alonso, D (2004) Molecular bases of the treatment of Alzheimer's disease with antioxidants: prevention of oxidative stress. *Molecular Aspects of Medicine*, 25, 17–123.

Wainwright, PE, Xing, HC, Mutsaers, L, McCutcheon, D and Kyle, D (1997) Arachidonic acid offsets the effects on mouse brain and behavior of a diet with a low (n–6):(n–3) ratio and very high levels of docosahexaenoic acid. *Journal of Nutrition*, 127, 184–193.

Wang, T and Hammond, EG (1999) Fractionation of soybean phospholipids by high-performance liquid chromatography with an evaporative light scattering detector. *J. Am. Oil Chem. Soc.*, *76*, 1313–1321.

Wang, T, Hammond, EG and Fehr, WR (1997) Phospholipid fatty acid composition and stereospecific distribution of soybeans with a wide range of fatty acid compositions. *J. Am. Oil Chem. Soc.*, 74, 1587–1594.

Wickremasinghe, RG, Porfiri, E and Hoffbrand, AV (1993) Evidence that inositol phospholipids, but not choline phospholipids, are a potential source of growth-regulating second messenger molecules in HL60 leukemia cells. *Leukemia Research*, 17, 763–9.

Wilson, TA, Meservey, CM and Nicolosi, RJ (1998) Soy lecithin reduces plasma lipoprotein cholesterol and early atherogenesis in hypercholesterolemic monkeys and hamsters: beyond linoleate. *Atherosclerosis*, 140, 147–153.

Zalejski, C, Zhang, Z, Quettier, A, Maldiney, R, Bonnet, M, Brault, M, Demandre, C, Miginiac, E, Rona, J, Sotta, B and Jeannette, E (2005) Diacylglycerol pyrophosphate is a second messenger of abscisic acid signaling in Arabidopsis thaliana suspension cells. *Plant Journal*, 42, 145–152.

Zeisel, SH (1992) Choline: an important nutrient in brain development, liver function and carcinogenesis. *American College of Nutrition*, 11, 473–81.

Zeisel, SH (2005) Choline, homocysteine and pregnancy. *American Journal of Clinical Nutrition*, 82, 719–720.

Zeisel, SH, Da Costa, KA, Franklin, PD, Alexander, E A, Lamont, JT, Sheard, NF and Beiser, A (1991) Choline, an essential nutrient for humans. *FASEB Journal*, 5, 2093–2098.

CHAPTER 2

Major sources, composition and processing

MICHAEL SCHNEIDER

LECITHOS – Functional Lipid Innovation and Consulting, Freinsheim, Germany

A. Introduction

Recent developments in food, nutritional and pharmaceutical applications of phospholipids have changed the perception that phospholipids are to be considered only as food additives (in the form of lecithin) or for the biochemical study of membrane phenomena. There is a dramatic increase in interest from industry, which has led to the development of new processes and new raw materials. The driving force behind this development is the nutritional industry (Szuhaj & Nieuwenhuyzen, 2003).

The food industry with its competitive need to diversify and differentiate has discovered phospholipids as a class of substances with synergistic functionality of emulsification and physiology. Many nutritional and clinical trials have been conducted to show the effects of phospholipids in prevention or treatment of certain disease conditions (see also Chapter 7). Because food is the driving force behind new developments, it is essential that products, raw materials and processes are all of food grade.

Present challenges for the industry are the topics of genetic modification (the European industry asks exclusively for non-GM ingredients) and the Novel Food issue. All new substances, ingredients, and processes designed for food use, which have not been in use prior to the Novel Food Regulation (1997) in a significant amount, have to go through a lengthy and costly approval procedure.

This chapter is confined to natural or naturally derived phospholipids and does not cover synthetic compounds.

B. Occurrence

The composition of natural phospholipid mixtures differs both in the proportion of the individual phospholipid present and in the composition of their component fatty acids. Commercially available phospholipids are extracted, concentrated or modified (chemically or enzymatically) from both animal and vegetable sources. They can also be obtained from fermented biomasses, but this has not yet been commercially applied, probably as the membrane (phospholipid) content of biomasses rarely exceeds 1% of dry matter. However, manipulation of the growing conditions of microorganisms could theoretically change both the class of specific phospholipids and also their content (Ratledge & Wilkinson, 1988).

Currently, the most prominent raw materials are vegetable oilseeds, such as soybeans, sunflower or rapeseed. Small volumes may be extracted from corn, rice bran and oat, but these do not play a significant role. Eggs provide the main animal source. Because of the interest in omega-3 fatty acids, marine raw materials are growing in importance (krill, fish roe, fish processing by-products) Also, cows' milk is of some interest because of its uniquely high sphingomyelin content.

Whereas vegetable phospholipids are mainly obtained as by-products from the industrial production and refining of vegetable oils, animal phospholipids are generally the result of specific processes exclusively designed to prepare the phospholipid products. This is one of the reasons why the prices for animal phospholipid mixtures are much higher than for their vegetable counterparts. It is important to mention that there is no natural source providing a single specific phospholipid type. This is because phospholipids are part of complex biological membrane systems or, in animals, part of the circulating lipoproteins, the emulsifying bile complex, or of lung surfactant. In all these biological systems, the concerted action of a number of different phospholipid classes is of crucial importance for optimal activity, maintenance, or performance of membrane-related biochemical processes.

Figure 2.1 shows the typical phospholipid double layer schematically. Polar head groups point both to the inner or outer water-based environment; fatty acids form a lipophilic inner domain and act as a barrier between inside and outside and also provide optimal membrane fluidity. Imbedded in the membrane structure are proteins, enzymes, receptors and ion channels.The phospholipid content of biomass is usually relatively small, with the exception of eggs and sperm, both providing all the necessary ingredients for a developing living organism. Table 2.1 gives details of the phospholipid content of a selected number of different raw materials.

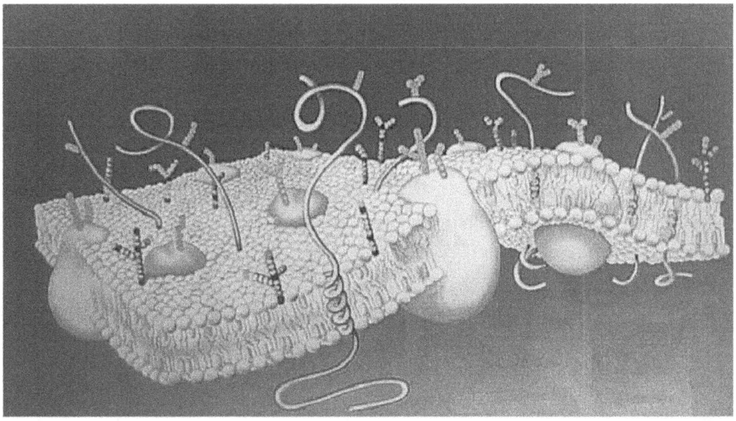

Figure 2.1. Model of a biological membrane (from *Scientific American*).

Table 2.1. Phospholipid content of biological materials (% of dry matter)

	Soybeans	Sunflower seeds	Rice bran	Egg yolk	Salmon roe	Milk
Phospholipid content	0.5	0.2	1	17	8	0.1

Table 2.2. Main phospholipid classes of commercial interest

Class	Abbreviation
Glycerophospholipids	
Phosphatidylcholine	PC
Phosphatidylethanolamine	PE
Phosphatidylserine	PS
Phosphatidylinositol	PI
Phosphatidylglycerol	PG
Diphosphatidylglycerol	DPG
Phosphatidic acid	PA
N-Acyl-phosphatidylethanolamine	NAPE
Sphingophospholipids	
Ceramide phosphocholine (sphingomyelin)	SPM

1. Phospholipid classes

Two main classes of natural phospholipids occur in quantities sufficient to

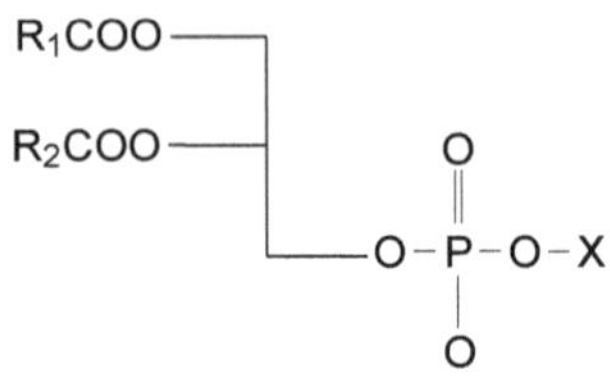

Figure 2.2. Structure of diacylglycerophospholipids; X is residue from choline (PC), ethanolamine (PE), inositol (PI), glycerol (PG), water (PA).

allow industrial manufacture and application: glycerophospholipids with a glycerol and sphingophospholipids with a sphingosine backbone (Chapter 1).

(a) Glycerophospholipids. These are derivatives of triacylglycerols with one fatty acid substituted by a phosphoric acid ester of an amino alcohol (e.g., choline, ethanolamine), a polyol (e.g., glycerol, inositol) or, in the case of phosphatidic acid, by free phosphoric acid (see Figure 2.2). All natural materials also contain some small amounts of mono-acylphospholipids, so-called lysophospholipids.

(b) Sphingophospholipids. In sphingomyelin (the most abundant sphingophospholipid), glycerol is replaced by a long-chain amino alcohol (sphingosine), linked to a fatty acid by an amide bond (called ceramide) and having the phosphoric acid ester of choline as head group (see Figure 2.3).

The phospholipid composition (Table 2.3) is a fingerprint typical of the raw material used, as is the composition of the fatty acids linked to the phospholipid backbone (Table 2.4). All figures in Table 2.4 reflect the fatty acid composition of natural phospholipid mixtures free of oil (triacylglycerols). Individual classes of phospholipids can differ considerably from the data in Table 2.4, and all phospholipids, even from a single source, are significantly different in their fatty acid profile (Gornall & Kuksis, 1971).

Natural polar lipid mixtures also contain another class of (phosphorus free) polar lipids, the glycolipids. Like phospholipids, glycolipids can be considered as derivatives of triacylglycerols in which one fatty acid is replaced by a carbohydrate molecule (e.g., digalactosyldiglyceride) or again as derivatives of sphingosine, such as lactosylceramide (see Chapter 8).

In vegetable oil processing, phospholipids are extracted, concentrated, isolated or modified starting from lecithin, which is obtained as a result of the first step of oil refining, the so-called degumming process (see Section C.1.a).

It is important to point out that in this chapter the term lecithin is used in the commercial sense, of a complex mixture of polar and neutral lipids, some

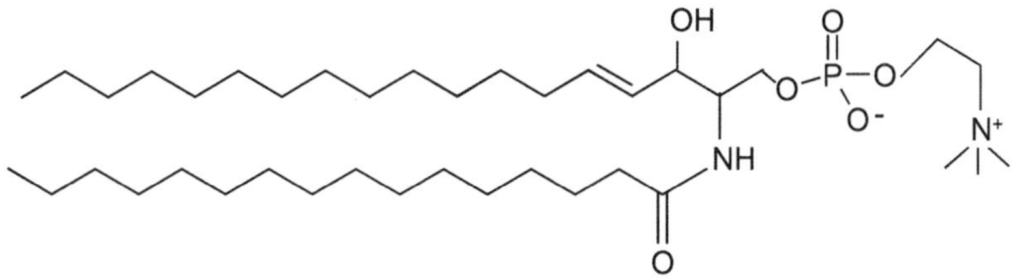

Figure 2.3. Structure of sphingomyelin.

Table 2.3. Phospholipid composition in oil-free polar lipid extracts from commercial raw materials (%)

	Soya	Rapeseed	Sunflower seed	Corn	Egg	Milk	Salmon roe
PC	22	25	25	30	74	27	86
PE	14	22	11	3	17	30	6
PI	12	15	19	16	1	2	2
PS			1	1	–	7	
PG/DPG	2		1	1	1		1
PA	6		3	9	–	1	
NAPE	3		1	3	–		
SPM					3	29	2
Lyso-PLs	2	19		5	2		2
Others	4	19			1	4	
Reference	author's data	Appleqvist & Ohlson, 1972	Grewal *et al.*, 1978	Weber, 1981	Palacios & Wang 1985	Kuksis, 1985	author's data

Where the phospholipids do not add up to 100%, the difference is made up of glycolipids and carbohydrates, especially for vegetable lecithins.

Table 2.4. Main fatty acid composition of phospholipid mixtures (%)

	Soya	Rapeseed	Sunflower seed	Corn	Egg	Milk	Salmon roe
16:0	20	18	15	22	30	22	17
18:0	4	1	3	2	16	13	2
18:1	12	21	13	27	29	36	18
18:2	57	48	69	48	14	12	2
18:3	7	7		1	1	1	1
20:4					5		1
20:5							17
22:6						3	31
Other		5			5		11
Reference	Weber, 1981	Hougen *et al*, 1985	Grewal *et al.*, 1978	Weber, 1981	Palacios & Wang 2005	Kuksis, 1985	author's data

glycolipids and carbohydrates (the legal and commercial terminology). The scientific literature sometimes uses the term lecithin for just one defined phospholipid, namely phosphatidylcholine.

C. Phospholipid processing

1. Vegetable phospholipids

(a) Fluid (crude) lecithins. The preferred route to crude vegetable phospholipids is to separate them as a by-product of the physical refining process from crude vegetable oils. The first step of refining, called degumming, is based on the fact that phospholipids tend to hydrate and to form high molecular aggregates in the presence of water. These are insoluble in oil and can be removed as wet gums by mechanical separation. After drying, the gums give lecithins with approximately 50 to 65% polar lipids, predominantly phospholipids.

There are many variations of the degumming process (Flider, 1985; Clausen, 2001), all aiming to reduce the remaining phosphorus in the oil to a minimum. Low residual phosphorus is of importance for further processing, particularly for hydrogenation, as the phosphorus compounds act as a catalyst poison. Degumming conditions have a strong influence on crude vegetable lecithin composition and quality (Table 2.5).

Based on such crude lecithins, the industry has built a complex system of

Table 2.5. Composition of a commercial fluid soya lecithin (%)

Phospholipids [a]	52
Phosphatidylcholine (PC)	15
Phosphatidylethanolamine (PE)	14
Phosphatidylinositol (PI)	12
Phosphatidic acid (PA)	4
N-Acyl-phosphatidylethanolamine (NAPE)	2
Other phospholipids (e.g., lyso-)	5
Glycolipids	6
Carbohydrates	4
Neutral lipids	38
Fatty acids	
Palmitic acid	18
Stearic acid	5
Oleic acid	11
Linoleic acid	59
Linolenic acid	7

From Nieuwenhuyzen (1976).

[a]Neutral lipids are also present and add to 100% along with the polar lipids and glycolipids.

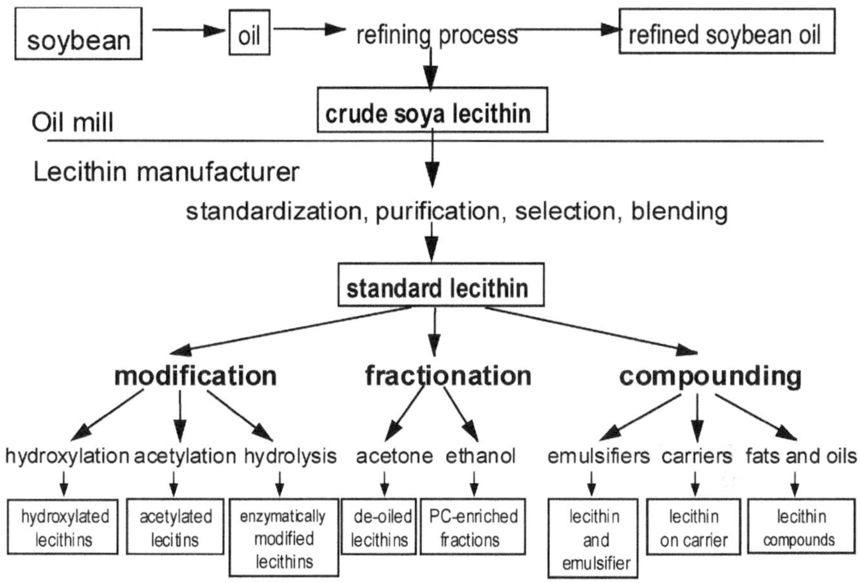

Figure 2.4. Vegetable lecithin processing.

standardizing, refining, fractionation and modification steps to serve the different industrial applications. Generally the products have properties optimized for their application.

The simplified scheme in Figure 2.4 illustrates the processing basics of vegetable phospholipids/lecithins, including fractionation and enzyme and chemical modifications. Fluid (crude) lecithins (approximately 250,000 tons annually) represent the biggest market share of all phospholipid-containing products. In general, they serve as non-toxic, but highly functional, food and feed emulsifiers, dispersants, structure modifiers and stabilizers. Soya dominates this segment with more than 80% market share. During the past few decades, the quality of fluid vegetable lecithins has improved dramatically, especially with regard to colour and impurities (so-called hexane insolubles, HI, which are soybean fines resulting from the oil extraction process). Manufacturers have tried to get highly filtered soya lecithin exempted from allergy labelling, but there are still minor traces of protein remaining in commercial lecithins.

(b) Solvent technologies. Commercial solvent technologies are limited to a small selection of solvents appropriate to the image and use of a natural functional ingredient for the various finished goods. Because of the outstanding importance of the food and nutritional industry, the solvents of choice are:

- hexane
- acetone
- ethanol
- compressed gases
- hexane, methanol, iso-propanol.

(i) *Membrane degumming.* It has long been known that non-polar solvents such as hexane do not dissolve polar lipids in a molecular dispersed manner, contrary to neutral lipids. Polar lipids, like phospholipids, form high molecular lamellar structures (Sen Gupta, 1986). Based on this observation, a new degumming process has been developed (Cargill, 2001). The hexane solution from oil seed extraction contains a mixture of polar lipid aggregates dispersed in a homogeneous solution of neutral lipids. Passing such solutions through ultrafiltration modules with cut-offs of approximately 20,000 Dalton leads to a transport of the neutral lipid solution across the membrane (permeate), whereas the polar lipid aggregates are retained and concentrated in the retentate. Operating such ultrafiltration processes correctly, water degumming can be avoided. Lecithin is separated in almost exactly the same quality and composition as with the standard water degumming process. The advantage for the oil mill is extremely low residual phosphorus in the degummed oil, lower than with conventional procedures.

Modern membrane systems are solvent-resistant, even against hexane, and allow long standing times. However, the limiting problem is membrane fouling caused even by minimal protein impurities in the extracted oil, even when it has undergone very careful filtration steps.

(ii) *Membrane de-oiling.* Exhaustive membrane separation of polar lipids (phospho- and glycolipids) from neutral lipids can lead to completely de-oiled polar lipid products (de-oiled lecithins) which are already on the market. However, the commercial product is not made by exhaustive degumming since the throughput volume of soybean extraction solutions is too high, causing serious fouling problems as mentioned. The commercial product is made by dissolving water-degummed lecithin in hexane, followed by membrane separation (Cargill, 2001; Archer-Daniels-Midland Company, 2000).

(iii) *Acetone de-oiling.* Vegetable lecithins contain approximately 30–40% neutral lipids, predominantly triacylglycerols. To improve the handling of the highly viscous crude lecithins and to improve dispersibility properties, industry makes use of the fact that polar phospho- and glycolipids are almost insoluble in acetone, whereas neutral lipids dissolve (Gober *et al.*, 1995).

Batch or continuous acetone extraction leads to powdered or granulated products containing only 2–3% residual neutral lipids. The products are slightly hygroscopic, but flow freely especially after the addition of minor amounts of flow agents such as tricalcium phosphate or silica.

Because acetone also removes most of the natural tocopherols (approximately 1,000 ppm in crude soybean lecithin) the shelf life of de-oiled products is usually limited to 12 months.

A drawback of using a highly reactive solvent like acetone for lecithin treatment is the formation of low amounts of reaction products. Acetone tends to dimerize to di-acetone alcohol and finally to form mesityl oxide ($Me_2C=CH$ $COCH_3$) by splitting off water. The formation of mesityl oxide is strongly dependent on the contact time between acetone and phospholipids and its typical cat urine smell can be recognized, especially in powder lecithins from which acetone and its reaction products have not been completely removed.

Another problem with the use of acetone for de-oiling is acetone's tendency to form Schiff base type adducts with the amino phospholipids such as phosphatidylethanolamine (Kuksis *et al.*, 2005). These observations, together with environmental concerns and the wish to avoid lots of oxygen (air) during drying of the usually highly unsaturated phospholipids have stimulated attempts to develop alternative procedures.

(iv) *Carbon dioxide de-oiling.* Compressed carbon dioxide shows excellent solubility for neutral lipids at moderate temperatures and high pressures and does not dissolve polar lipids at all. For more than 25 years, several companies from both the lecithin and the equipment industries have tried to transfer laboratory (batch) results into industrial scale, so far without success.

The problem to overcome was the dramatic increase in the viscosity of lecithin during successive removal of oil. Diffusion of fresh CO_2 into the mass and movement of CO_2 out of the mass was too slow, so that the process took hours and even then led to non-homogeneous products. Poor diffusion could be solved with a clever spray nebulizing approach, creating small lecithin droplets, as in spray drying, which could be de-oiled more or less instantaneously (Lucas Meyer, 1994).

The problem is not the separation of the oil from the polar lipids, but the removal of the oil-free powders continuously from the pressurized containers, sometimes at more than 500 bar. This process could only compete with acetone technology if it proved possible to operate CO_2 processing continuously. Approaches with mixtures of propane and CO_2 or even pure propane have been more successful in moving towards continuous processing, but this has not yet been used in industrial processing (Peter, 1994).

Table 2.6. Typical composition of a commercial de-oiled soybean lecithin (%)

Phospholipids	81
Phosphatidylcholine (PC)	24
Phosphatidylethanolamine (PE)	22
Phosphatidylinositol (PI)	14
Phosphatidic acid (PA)	8
N-Acyl-phosphatidylethanolamine (NAPE)	6
Lysophospholipids	3
Others	4
Glycolipids	10
Carbohydrates	6
Neutral lipids	3
Triglycerides	2
Free fatty acids	0.5
Sterols	traces
Moisture	1

The advantages of acetone-free de-oiling are manifold: low temperature operation, oxygen-free process, no reaction between acetone and aminophospholipids, environmentally friendly process. Table 2.6 shows the composition of a commercial de-oiled soybean lecithin.

(v) *Alcohol fractionation.* The industrial production of vegetable lecithin 'fractions' mainly uses ethanol or ethanol/water systems. The principle is that phosphatidylcholine is preferentially soluble in ethanol, whereas phosphatidylethanolamine (PE) distributes more or less evenly between the soluble and the insoluble phase. The more acidic phospholipids such as phosphatidylinositol (PI) and phosphatidic acid (PA) remain practically insoluble. Commercial interest in both these fractions arises from the difference in their technological functionality (the soluble fraction is an o/w emulsifier and the insoluble fraction a w/o emulsifier). The interest of the industry in PC enriched products for dietetic purposes is also a driving force.

Alcohol fractionation can be carried out both before and after de-oiling. Table 2.7 shows the changes in phosphatidylcholine concentration after fractionation of both a liquid and a de-oiled soybean lecithin (Schneider, 1989). For fluid lecithin fractionation especially, continuous processes with either counter-current or concurrent extraction principles have been applied.

The methods described for soybean lecithin can in principle be applied also to other crude vegetable phospholipid products (lecithins). The results are similar, but not used commercially.

Table 2.7. Alcohol fractionation of soybean lecithin

	Triacylglycerols (%)	Phosphatidylcholine (%)	Yield (%)
Crude lecithin	39	15	
Soluble fraction:			
Methanol	25	39	23
Ethanol (100%)	60	22	26
Ethanol (95%)	48	33	15
Ethanol (90%)	23	46	12
Isopropanol	54	19	49
Insoluble fraction:			
Ethanol (95%)	38	7	85
De-oiled lecithin	2.5	24	
Soluble fraction:			
Ethanol (100%)	2.5	52	28
Insoluble fraction:			
Ethanol (100%)	2.5	12	72

(vi) *Chromatographic purification.* Schneider (1989) has reviewed this topic. Generally, chromatographic separation is the method of choice for the preparation of polar lipids of high purity. It is almost impossible to give a short summary of the numerous methods published in the scientific and patent literature. Most of them are used for the quantitative analysis of tissue extracts rather than for the production of commercial quantities of phospholipids or glycolipids.

The best separation results are usually achieved by combining different separation techniques; for example, solvent extraction of crude lipid mixtures in combination with adsorption processes. Among the long list of different adsorbents are aluminum oxide, silica, Florisil (magnesium silicate), DEAE (diethylaminoethyl)-cellulose and ion-exchange resins (Table 2.8). Alumina and silica are of great importance in industrial applications. Both materials are used in low- and medium-pressure chromatography to produce large quantities of soybean- and egg PC concentrates or isolates.

A typical alumina separation is as follows: de-oiled soybean lecithin is first fractionated with ethanol to produce a soluble fraction with about 50% PC. This solution is then passed down a column filled with aluminium oxide. Nearly all of the PC accompanying phospholipids are bound to the alumina and the effluent consists of almost pure choline phospholipids. One drawback of the method is the impossibility of recycling the adsorbent and removing lysophosphatidylcholine (LPC) from the material.

Applying alumina chromatography to egg phospholipids shows that, with ethanol as solvent, sphingomyelin has the same elution behaviour as PC and is included in the PC/LPC effluent.

Table 2.8. Chromatographic systems for phospholipid fractionation/purification

Adsorbent	Solvent system
Aluminum oxide	Ethanol
	Chloroform/methanol
Silica gel	Hexane
	Ethanol
	Hexane/isopropanol/water
	Chloroform/methanol
	Acetone/chloroform/methanol
	Chloroform/acetone
Florisil	Chloroform/methanol
Amberlite	Ethanol
	Ethylacetate/ethanol
DEAE-Cellulose	Chloroform/methanol/acetic acid

The separation of highly purified PC as well as other phospholipids is possible using stepwise elution with a chloroform/methanol/water mixture of different composition.

Compared to the alumina separation, silica chromatography seems to be more flexible with respect to solvent systems, separation characteristics and reconditioning of the column.

Silica is also used for selective elution processes to obtain the different polar lipid classes of a given lecithin. The preferred solvent systems are chloroform/methanol and methanol/acetone mixtures. Separation is mainly done stepwise, but by gradient elution in some cases. For more details about the wide range of possibilities for chromatographic separation, the reader is referred to the literature (Schneider, 1989).

Today's industrial scale chromatography is mainly focused on the preparation of PC from egg or soya. Those are of great commercial interest and available even in ton quantities.

Preparation of other phospholipids such as PS and PI is of growing interest. The chosen separation technique is large-scale HPLC, which is also used for the production of semi-synthetic and synthetic phospholipids. However, because of cost considerations, more and more enzymatic processes have been developed leading to sufficiently pure phospholipid concentrates (see Chapter 3).

Even purified PC from a conventional chromatographic process is still a mixture of various molecular species differing in fatty acid composition (Kuksis, 1967).

The literature gives little indication of phospholipid separation on the basis of the degree of unsaturation. For analytical purposes, the use of argentation

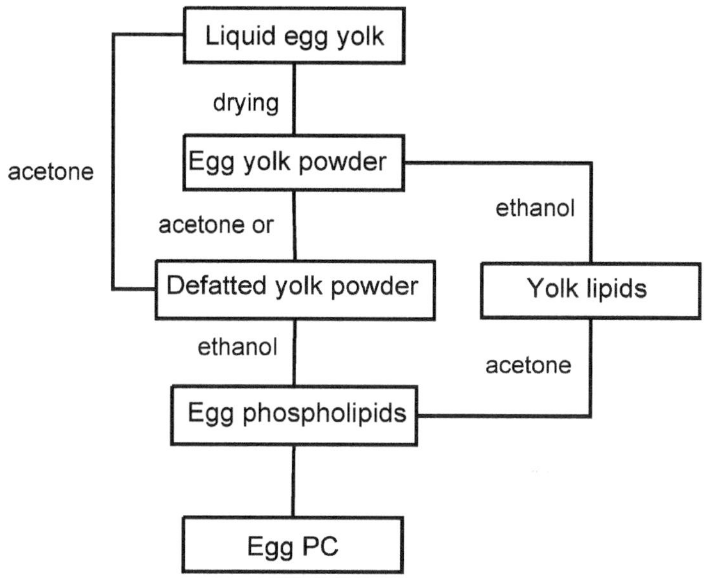

Figure 2.5. Egg yolk phospholipid processing principles.

chromatography with silver nitrate impregnated silica as an adsorbent is described. The higher the degree of unsaturation, the more firmly the phospholipids bind to the stationary phase and can be separated from each other.

A more recent development in this area is to use reverse phase HPLC. However, it is not known that these processes are yet used commercially. To synthesize the desired molecule seems to be more efficient (see Chapter 3).

2. *Animal phospholipids*

In contrast to vegetable phospholipids there is no large-scale industrial animal oil production and therefore phospholipids have to be specifically extracted using the various technologies.

(a) Egg phospholipids. Egg yolk is probably the richest phospholipid source in nature. Liquid chicken egg yolk contains 30% lipids (60% of dry matter), with approximately 9% phospholipids (18% of dry matter) (Juneja, 1997)

Among animal derived phospholipids, egg phospholipids are currently the products of greatest commercial interest (emulsifier in pharmaceuticals, intravenous fat emulsions, infant formula). The commonest processing alternatives are summarized in Figure 2.5. Other variations of extraction/purification are also described in the (patent) literature (Nielsen, 1980).

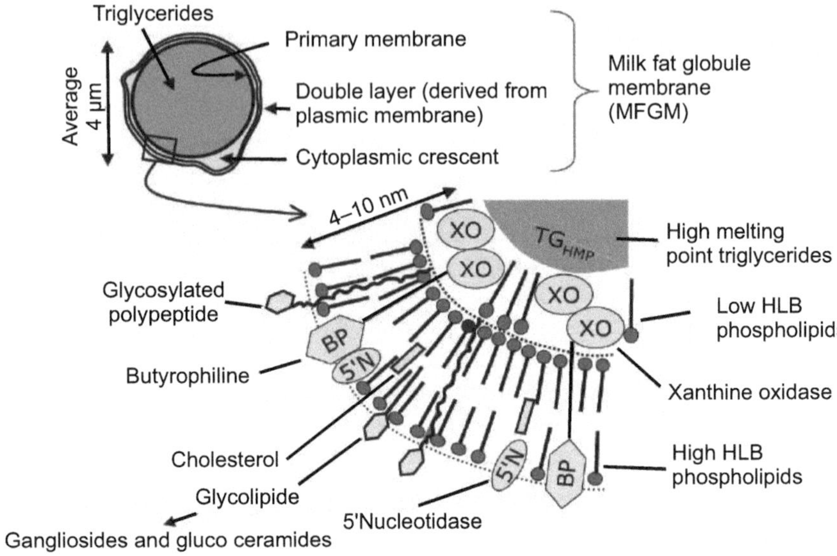

Figure 2.6. Milk fat globule membrane.

(b) Milk phospholipids. The phospholipids in milk are found in the milk fat globule membrane, a very complex structure surrounding and stabilizing the fat droplets (Figure 2.6). In liquid cows' milk, the phospholipid content is only about 0.1%.

Methods for milk phospholipid extraction and purification are numerous (Crane & Horall, 1943; Cerbulis, 1967). However, the simplest starts from dry, ultrafiltered butter serum (to remove lactose) (Arla Foods, 2006). On a dry basis, this product already contains approximately 20% phospholipids. After ethanol and acetone treatment, purified milk phospholipids are obtained.

Milk phospholipids offer a unique phospholipid profile with an almost equal content of PC, PE and sphingomyelin (SPM), and substantial amounts of natural phosphatidylserine (PS) (see Table 2.3). Milk phospholipids are attractive compounds for cosmetic applications and particularly liposome formulations. This is because of their excellent emulsifying properties, their resistance against oxidation (only 15% of the fatty acids are polyunsaturated) and a phase transition temperature of liposomal membranes close to body temperature (28°C).

Milk phospholipids are also attractive from a nutritional point of view. Milk PS can improve cognitive performance and stress resistance and can help improve gut health.

Table 2.9. Lipid composition of commercial marine phospholipid products (%)

	Krill	Fish processing by-product	Salmon eggs
Neutral lipids	65	40	65
Phospholipids	35	60	35
Phoshatidylcholine	30	23	30
Phosphatidylethanolamine	2	17	2
Phosphatidylinositol	2	12	2
Phosphatidylserine	–	4	–
Sphingomyelin	1	4	1
LC-PUFA			
EPA	15	10	17
DHA	9	20	31

(c) Marine phospholipids. Since Bang and co-workers (1976) discovered the health benefits of long-chain polyunsaturated fatty acids (LC-PUFA) of the omega-3 variety, there has been much enthusiastic interest in omega-3 lipids, with new benefits described all the time. It has also been discovered that the ratio of omega-6 to omega-3 fatty acids in modern food is heavily imbalanced compared with historic ratios when humans were hunter-gatherers. These findings have led to many clinical trials and observational reports on the benefits of LC-PUFAs, and there are now a large number of fish oils and fish oil concentrates on the market.

However, there are a number of problems with neutral lipids from fish: they are highly sensitive to oxidation, with rapid development of a fishy smell and taste, and slow loading of target structures in the human body with LC-PUFAs.

Marine phospholipids have many advantages over the neutral lipids. They are much more resistant to oxidation, provide LC-PUFAs with much better bioavailability and seem to have a broader spectrum of health benefits. Marine phospholipids are extracted from different raw materials (De Koning, 2003). Currently, the leading product on the market is a krill extract with approximately 35% phospholipids (Neptune, 2001). There is also a product made from fish processing by-products and another from salmon eggs (roe) (Table 2.9). Extraction and processing methods are similar to those for egg yolk; for example, using combinations of ethanol and acetone processing starting from wet or dried biomass (see Figure 2.5 for egg yolk processing). The overall composition of egg yolk and fish roe are very similar; both are extremely high in phospholipids and similar in composition by lipid classes.

(d) Brain phospholipids. Brain phospholipid extraction is usually based on various modifications of the Folch procedure (Folch *et al.*, 1951, 1957). Removal of non-lipid impurities is a lengthy process requiring further solvent extraction or gel chromatography. A modification of the extraction process, which does not require extraction or adsorption steps, is described in Phillips and Privett (1979). Industrially, brain phospholipids lost their market because of the BSE problem and were taken from the market in the mid 1990s.

D. Modification of phospholipids

The chemical and physical properties of phospholipids depend on their molecular structures. To meet industrial requirements for specific application profiles, hydrolysis, hydroxylation, acetylation, hydrogenation and to a large extent enzymatic processes have been applied to modify the original structure of the natural phospholipids, and consequently their physicochemical and nutritional properties.

1. Chemical modification

There are various ways of modifying phospholipid molecules chemically (Schmidt, 1985), but only a few of them are used commercially, mainly because none of the resulting products has food grade status in Europe. In the US, hydroxylation is allowed for baking applications and acetylation for release products.

(a) Chemical hydrolysis. Food laws do not allow chemical hydrolysis of phospholipids (lecithins). However, there is one product on the market as a supplement in the US and Far East made by controlled hydrolytic cleavage of both fatty acids from phosphatidylcholine with ethanolic potassium hydroxide (Istituto Chemiotherapico, 1985). The resulting glycerophosphocholine (GPC) has shown interesting effects on cognitive and physical performance as well as stimulating human growth hormone release.

(b) Acetylation. See Chapter 6, Section B.

(c) Hydroxyation. See Chapter 6, Section C.

(d) Hydrogenation. All natural phospholipid mixtures (lecithins) contain more or less high levels of unsaturated fatty acids (see Table 2.4). This makes all products sensitive to oxidative processes, which finally lead to rancidity and an unpleasant taste and smell. During catalytic hydrogenation of natural

Table 2.10. Commercial phospholipid products and their predominant use

Product	Application
Fluid soybean lecithin (oil containing)	AF/FO/I/CO/PH/DI
Ethanol soluble fractions (oil containing)	FO/CO/PH/DI
Ethanol insoluble fractions (oil containing)	AF/FO
Partially hydrolysed fluid lecithin	FO
Hydroxylated fluid lecithin	AF/CO/I/(FO)
Sunflower seed lecithin	AF/FO
Rapeseed lecithin	AF/FO/I
Corn lecithin	AF/TI
De-oiled soybean lecithin (powder, granules)	AF/FO/I/CO/PH/DI
De-oiled soybean lecithin fractions (PC 50%–80%)	FO/CO/PH/DI
Purified soybean PC (>90%)	PH/CO
Egg lipid mixes (PL containing)	DI/PH
Egg phospholipids (de-oiled)	FO/DI/CO
Milk phospholipids	CO/PH
Fish egg phospholipids	DI
Krill phospholipids	DI
Brain phospholipids	PH
Head group exchanged phospholipids (PS)	PH/DI

FO: food; AF: animal feed; I: industrial; CO: cosmetics; PH: pharma; DI: dietetics.

phospholipids, the unsaturated fatty acids become more saturated. As a result, the products are more stable against oxidation, but at the same time less soluble in fats and oils. Attempts to produce partially hydrogenated phospholipids failed, as it is very difficult to terminate hydrogenation reproducibly at a defined point of hydrogen uptake. Hydrogenation so far is limited to de-oiled soya lecithin and pure phosphatidylcholine. Hydrogenation of oil-containing lecithins failed, as the oil portion gets hydrogenated very quickly and the lecithin itself does not react. As a side effect of hydrogenation, most of the natural pigments are destroyed, turning the resulting products into white to off-white free-flowing powders.

The hydrogenation process for phospholipids requires quite drastic conditions regarding hydrogen pressure as well as type and quantity of catalyst. Only palladium and platinum catalysts are resistant enough against the poisoning effect of acidic phospholipids and sulfur-containing traces of impurities to allow reasonable conversion rates (American Lecithin, 1954; Sokolski & Omaraliev, 1968). The lack of food grade status and the high cost for a process requiring solvent, high pressure, and high catalyst quantity has limited the

Table 2.11. Major companies supplying a range of phospholipid preparations

Company	Website	Country	Fluid soya lecithin	Modified fluid lecithin	Other vegetable lecithins	De-oiled soya lecithin	PC fluid (35%)	PC de-oiled (>50%)	Hydro-genated PC	Soya PS	Egg PS	Milk PLs	Marine PLs	Synthetic PLs	Research PLs
ADM	www.admworld.com	US	x	x		x									
Solae	www.solae.com	US	x	x		x	x								
Cargill	www.cargillfoods.com	Germany	x	x		x	x	x			x				
Lipoid	www.lipoid.com	Germany						x	x	x	x			x	
Phospho-lipid	www.phospholipid.de	Germany					x	x	x						
Chemi	www.chemi.com	Italy							x					x	
Lecico	www.lecico.de	Germany	x	x	x	x	x			x	x	x	x		
Tsuji Oil Mill	www.tsuji-seiyu.co.jp	Japan	x			x	x	x							
Fresenius-Kabi	www.fresenius-kabi.com	Sweden									x				
Doosan	www.doosan.com	Korea						x		x	x			x	
Nippon Oils and Fats	www.phospholipid.jp	Japan								x					
Biofer	www.biofer.com	Italy						x					x		
Avanti Polar	www.avantilipids.com	USA						x						x	x
Belovo	www.belovo.com	Belgium									x				
Enzymotec	www.enzymotec.com	Israel						x		x			x		
Lipogen	www.lipogen.co.il	Israel								x					
Neptune	www.neptunebio-biotech.com	Canada											x		

application of hydrogenated phospholipids to pharmaceuticals and certain cosmetic applications.

2. *Enzymatic modification*

See Chapter 3 for a discussion of this topic.

E. Commercial uses

Commercial uses of phospholipids are summarized in Table 2.10 and details of suppliers and products are summarized in Table 2.11.

References

American Lecithin (1954) Hydrogenation of phosphatides. USP 2870179.

Appleqvist, LA and Ohlson, R (eds) (1972) *Rapeseed: Cultivation, Composition, Processing and Utilization*, Elsevier, Amsterdam.

Archer-Daniels-Midland Company (2000) Process for producing deoiled phospholipids. USP 61405190.

Arla Foods (2006) Phosphatidylserine enriched milk fractions for the formulation of functional foods. WO/2006/128465.

Bang, HO, Dyerberg, J and Hjorne, N (1976) The composition of food consumed by Greenland Eskimos. *Acta Medica Scandinavia*, 200, 69–73.

Cargill (2001), Method for removing phospholipids from vegetable oil miscella, method for conditioning a polymeric microfiltration membrane and membrane. USP 6207209.

Cerbulis, J (1967) Distribution of lipids in various fractions of cows milk. *Journal of Agricultural Chemistry*, 15, 784.

Clausen, K (2001) Enzymatic oil-degumming by novel microbial phospholipase. *European Journal of Lipid Science and Technology*, 103, 333–340.

Crane, JC and Horall, BE (1943) Phospholipids in dairy products. *Journal of Dairy Science*, 26, 935.

De Koning, AJ, (2003) Phospholipids of marine origin: A review of research in southern Africa. *South African Journal of Science*, 99, 521–525.

Flider, FJ (1985) The manufacture of soybean lecithins. In: *Lecithins* (BF Szuhaj and GR List, eds), AOCS Press, Champaign, pp. 21–37.

Folch, J, Ascoli, M, Lees, M, Meath, JA and Le Baron, FN (1951) Preparation of lipid extract from brain tissue. *Journal of Biological Chemistry*, 191, 833.

Folch, J, Lees, M and Stanley, GH (1957) A simple method for the isolation and purification of total lipids from animal tissue. *Journal of Biological Chemistry*, 226, 497.

Gober, KH, Guenther, BR and Lunebach, EM (1995) Isolation and analysis of phospholipids and phospholipid derivatives. In: *Phospholipids Handbook* (G Cevc, ed), Marcel Dekker, New York, pp. 1–22.

Gornall, DA and Kuksis, A (1971) Molecular species of glycero phosphatides and triglycerides of egg yolk lipoproteins. *Canadian Journal of Biochemistry*, 49, 51.

Grewal, SS, Sukhija, PS and Bhatia, LS (1978) Polar lipid composition during sunflower *Helianthus annuus* seed development. *Biochemie, Physiologie der Pflanzen*, 173, 11.

Hougen, FW, Thompson, VJ and Daun, JK (1985) Rapeseed lecithin. In: *Lecithins* (BF Szuhaj and GR List, eds), AOCS Press, Champaign, pp. 79–96.

Istituto Chemiotherapico (1985) Process for the preparation of L-alpha-glycerylphosphorylcholine from crude or deoleated lecithins. EP 217765.

Juneja, LR (1997) Egg yolk lipids. In: *Hen Eggs: Their Basic and Applied Science* (T Yamamoto, LR Juneja, H Hatta and M Kim, eds) CRC Press, Boca Raton, pp. 73–98.

Kuksis, A (1967) Determination of the complete structure of natural lecithins. *Lipids*, 2, 217.

Kuksis, A (1985) Animal lecithins. In: *Lecithins* (BF Szuhaj and GR List, eds), AOCS Press, Champaign, pp. 105–162.

Kuksis, A, Ravandi, A and Schneider, M (2005) Covalent binding of acetone to aminophospholipids in vitro and in vivo. *Annals New York Academy of Science*, 1043, 417–439.

Lucas Meyer (1994) Liquid high pressure spraying extraction process. EP 0785814.

Neptune Technologies & Bioresources (2001) Natural phospholipids of marine origin containing flavonoids and polyunsaturated phospholipids and their uses. EP 1417211.

Nielsen, JR (1980) A simple chromatographic method for purification of egg lecithin. *Lipids*, 15, 481–484.

Nieuwenhuyzen, WV (1976) Lecithin production and properties. *Journal of American Oil Chemists' Society*, 53, 425–427.

Novel Foods Regulation (1997) Regulation EC 258/97.

Palacios, LE and Wang, T (2005) Egg-yolk lipid fractionation and lecithin characterisation. *Journal American Oil Chemists Society*, 82, 571–578.

Peter, S (1994) Process for removing oil from raw lecithin. WO/1994/001004.

Phillips, F and Privett, OS (1979) A simplified procedure for the quantitative extraction of lipids from brain tissue. *Lipids*, 14, 590.

Ratledge, C and Wilkinson, SG (1988) *Microbial Lipids*, Academic Press, London.

Schneider, M (1989) Fractionation and purification of lecithin. In: *Lecithins, Sources, Manufacture and Uses* (BF Szuhaj, ed), AOCS Press, Champaign, pp. 109–130.

Schneider, M, personal data.

Sen Gupta, AK (1986) Novel developments in refining of edible oils. *Fette, Seifen, Anstrichmittel*, 88, 79–86.

Sokolski, DV and Omaraliev, T (1968) Hydrogenation of the lecithin fraction of soybean oil phosphatides. *Chemical Abstracts*, 68, 96992.

Szuhaj, BF and Nieuwenhuyzen, WV (2003) *Nutrition and Biochemistry of Phospholipids*, AOCS Press, Champaign.

Unilever (2000) Food composition suitable for shallow frying comprising sunflower lecithin. EP 1343385.

Weber, EJ (1981) Composition of commercial corn and soybean lecithins. *Journal of American Oil Chemists' Society*, 58, 898.

CHAPTER 3

Enzymatic modification of phospholipids and related polar lipids

XUEBING XU[1]*, ANDERS FALK VIKBJERG[2], ZHENG GUO[1], LONG ZHANG[1] AND ASHISH KUMAR ACHARYA[1]

[1]BioCentrum-DTU and [2]LiPlasome Pharma A/S, Technical University of Denmark, Lyngby, Denmark. * *Current address: Professor in Agro-Biotechnology Science, Department of Molecular Biology, University of Aarhus, DK-8000 Aarhus C, Denmark*

A. Introduction

Phospholipids (PLs) are a group of polar lipids containing one or more phosphate groups. PLs are amphiphilic in nature; that is, each molecule consists of a hydrophilic (water-loving) portion and a hydrophobic (water-hating) portion. Due to this character and their insolubility in water, PLs are ideal compounds for forming biological membranes. In industry, PLs are widely used as emulsifiers and surfactants on the basis of their amphiphilic nature.

There are two major classes of PLs, depending on whether they are based on glycerol or on sphingosine. Both are present in biological membranes. PLs based on glycerol are called glycerophospholipids and are the most abundant class of natural phospholipids. These are mainly phosphatidylcholines (PC), phosphatidylethanolamines (PE), phosphatidylserines (PS), phosphatidylinositols (PI), phosphatidylglycerols (PG) and cardiolipins. The structural diversity within each type of glycerophospholipid (and hence the use of the plural nouns) is due to the variability of the chain length and the degree of saturation of the fatty acid ester groups. Sphingomyelin is the major sphingosine-containing phospholipid and contains a fatty acid attached to sphingosine by an amide linkage.

Glycolipids are the major class of related polar lipids present in living things, forming integral and important parts of the cell constituents of plants, animals, fungi and bacteria. The simplest definition of glycolipids can be given as glycosides of acylated glycerol, sterol or ether-lipids. When it involves the glycosidic attachment of a saccharide (mono-, di- or tri-) moiety to mono- or diacylglycerols, the resulting lipid is called a glycoglycerolipid. It may have one or more sugar residues attached glycosidically to the free hydroxyl group of a mono- or diacylglycerol. In comparison with glycerophospholipids, the polar part of the lipids is changed from phosphate to a saccharide moiety as polar head group. This group of polar lipids is also included in this chapter.

Polar lipids, in particular phospholipids, have received much attention in academia and in many industries as multifunctional products. The modification of polar lipids to improve their function and applications is obvious. This is demonstrated in part in Chapter 6 concerning the chemical modification of PLs. This chapter examines the use of enzymes for the modification of phospholipids and other related polar lipids, and describes the general status of research progress and product possibilities, as well as the reaction strategies involved.

The use of enzymes for PL modification has moved quickly in recent years, not only in academic research but also in industry. Phospholipid hydrolysis to produce lyso-PLs has been developed using enzymes for the process in a few companies such as American Lecithin Company (www.americanlecithin.com), Cargill (www.cargilltexturizing.com), and Taiyo (www.taiyointernational.com). A few companies have also developed enzymes and technologies to exploit the enzymatic hydrolysis of PLs for the degumming of oils and fats, such as Novozymes (www.novozymes.com) and Diversa (www.diversa.com). Using enzyme technology, other modified PLs have also been marketed (www.enzymotec.com). Enzymatic modification of glycolipids in industry has not been reported so far; however, potential applications of glycolipids have pushed their industrial use in different commercial products (www.lipid.se). The fast growing use of enzymes for polar lipid modification arises from factors such as milder reaction conditions, less environmental pollution, better specificity for improved quality, and higher efficiency of reactions. The economic balance of the processes is, however, the key factor for industrial operations, and there has been a steady advance in enzyme production and enzyme technology in the past few years. This trend is expected to accelerate, and there will be a huge potential for wide-ranging applications in product and process development in the near future.

Different enzymes can be used for different modification purposes. As illustrated in Figure 3.1, a number of PL ester bonds can be modified with appropriate enzymes. For acyl modifications, the natural enzymes will be phospholipase A_1 (EC 3.1.1.32) and A_2 (EC 3.1.1.4) (PLA_1 and PLA_2, respectively). However, lipases (EC 3.1.1.3) in general can also be used for such purposes, in particularly for modification at the *sn*-1 position. No lipase is

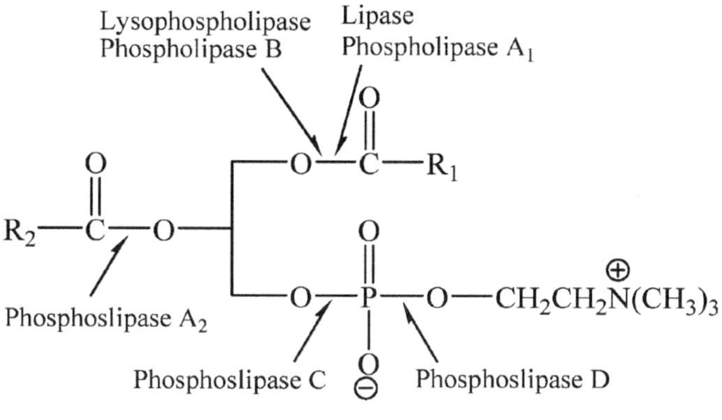

Figure 3.1. Specificity of phospholipases and lipases towards the phospholipid structure.

known to react selectively at the *sn*-2 position. Only PLA_2 in the phospholipase category has such specificity. Phospholipase D (EC 3.1.4.4) (PLD) is the only potential enzyme for polar group modification. Phospholipase C (EC 3.1.4.3) (PLC) can hydrolyse the bond between the glycerol OH group and the phosphate group but, unfortunately, the enzymes in this category are, at the moment, only active for the hydrolysis reaction. For glycolipids, lipases can be used also for acyl group modification. For the modification of saccharide group, the type of enzymes is diverse, depending highly on the nature of the saccharide group. Transglycosidases (EC 2.4), glucosidases (EC 3.2.1), galactosidases (EC 3.2.1), and even cellulases (EC 3.2.1.4) can be used for the reactions with the glycosidic bonds.

B. Enzymatic production of lysophospholipids

Lyso-PLs are partially hydrolysed PLs containing one fatty acid. Partially hydrolysed lecithins possess improved emulsifying properties compared to natural lecithin (Schneider, 1997). The hydrophilic/lipophilic balance (HLB) of the PL mixture increases with the degree of hydrolysis. PLs then become more hydrophilic and thus increase the ability to form oil-in-water emulsions. With a higher degree of hydrolysis smaller droplets are usually also generated in a comparative emulsion process, which gives more stable emulsions (Bueschelberger, 2004).

Different enzymes (e.g., acyl hydrolases) can be used for PL acyl hydrolysis. Although widespread in nature, only a few such enzymes have become available in large quantities for industrial purposes. Hydrolysis of PLs with PLA_1 (and with *sn*-1,3 specific lipases) yields 2-acyl lyso-PLs, and hydrolysis with PLA_2 yields 1-acyl lyso-PLs. Microbial lipases have been the most attractive enzymes for the modification of PLs in the *sn*-1 position since they are commercially accessible. Phospholipase B (PLB) can hydrolyse both positional ester bonds; however, it has not been used for PL modification, mainly because of its broad substrate selectivity.

For some years, industrial application of enzymes for lecithin modification by partial hydrolysis has mostly used PLA_2 from porcine pancreas (Chung & Yang, 2004; Colarow *et al.*, 1994; Buikstra *et al.*, 1997; Yesair, 1998). Enzymatic hydrolysis of phospholipids to lyso-PLs has also been applied during the degumming step in the refining of vegetable oil. Enzymatically generated deacylated PLs are more water soluble, and thus facilitate wash-out. Recently, a microbial PLA_1 has been introduced to industry for replacement of porcine pancreatic PLA_2 for the degumming process (Clausen, 2001).

These enzymes can also be used to produce lyso-PLs by alternative methods such as alcoholysis or esterification of glycerophosphorylcholine and transacylation of 2-acyl lyso-PLs to 1-acyl PLs.

C. Enzymatic modification of fatty acid profiles of glycerophospholipids

Natural phospholipids contain a variety of fatty acids, the proportions of which depend on the source. In some cases, it is desirable to have well-defined PL products with specified fatty acids in one or both positions. By replacing the existing fatty acids in an original phospholipid molecule with desired ones, new physical properties and special functions can be achieved (Huang, 2001). For example, it has been reported that the exchange of long-chain polyunsaturated fatty acid (PUFA) with short- or medium-chain fatty acids in PLs may improve their heat stability (Chmiel *et al.*, 1999). Medium-chain or long-chain saturated fatty acids may be incorporated to modify emulsification properties, to modify physiological value or to improve oxidation stability of the PL (Pedersen, 2001). PUFA, such as linoleic, arachidonic, α-linolenic, eicosapentaenoic, docosahexaenoic or γ-linolenic acids may be incorporated to improve the physiological or nutritional value of PLs. PUFA at the secondary position (*sn*-2 position) of a PL molecule are especially bioeffective (Takahashi & Hosokawa, 2001). Phospholipids containing docosahexaenoic acid (DHA) may have potential medical applications in promoting cell differentiation in leukaemia, enhancing survival of tumour bearing mice, and preventing cerebral apoplexy (Eibl & Unger, 1988; Jenski *et al.*, 1995; Lasson-Backstrom, 1995; Sakai *et al.*, 1992; Zerouga *et al.*, 2002).

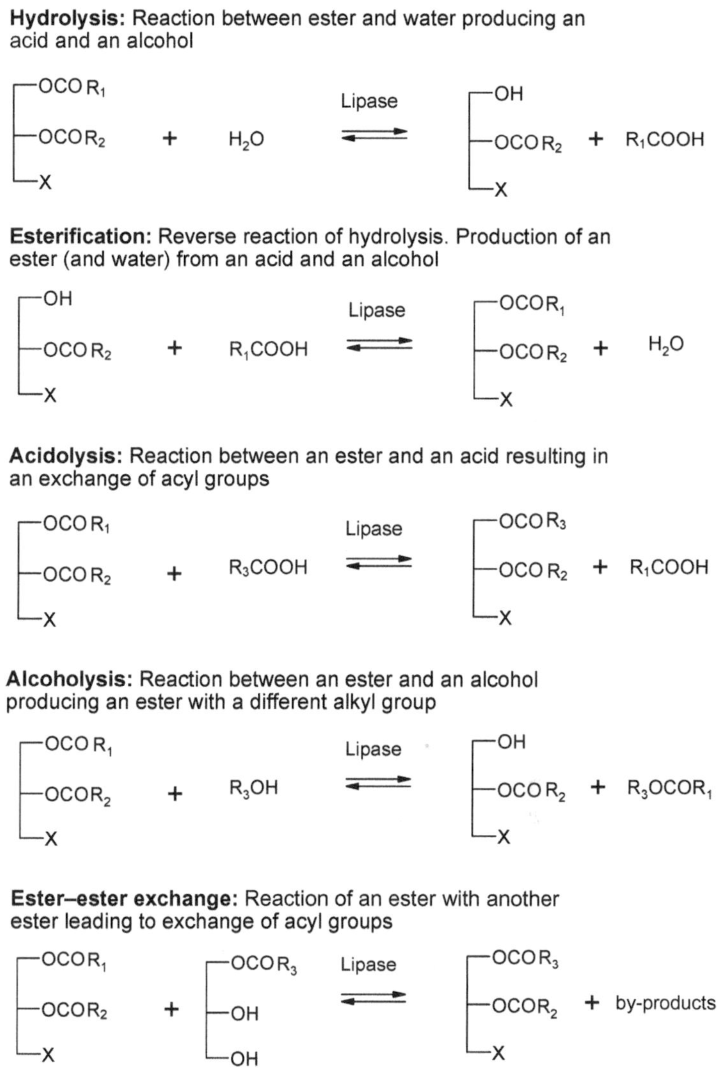

Figure 3.2. Reactions catalysed by lipase for phospholipid modification.

1. Reaction strategies for the modification of acyl groups

Many hydrolytic enzymes can also catalyse ester bond formation. With appropriate available enzymes, the manipulation of PL structure can be complicated but versatile. Various reaction routes can be implemented and have been used for the production of different products. Possible reactions, which can be catalysed by lipases and phospholipases, are presented in Figure 3.2. Reaction schemes are only illustrated for lipase-catalysed reactions; however, similar reactions can be performed with PLA_2 for modification of fatty acids at the *sn*-2 position. In the literature, acidolysis, alcoholysis and ester–ester exchange are collectively referred to as interesterification or transesterification.

Production of diacyl-PLs with modified fatty acid composition (structured PLs) can be achieved in one step by acidolysis or ester–ester exchange. Alternatively, reactions are performed in two steps by first removing fatty acids by hydrolysis or alcoholysis, followed by re-esterification of the lyso-PLs. The nature of the final product will mainly determine the reaction route (Adlercreutz *et al.*, 2003). In the one-step reactions, the fatty acid composition of the product will be a mixture of the original fatty acids and those being incorporated. In the two-step process, it is possible to isolate the lyso-PLs before the subsequent esterification reaction, which gives rise to PLs with high purity. Compared to the one-step process, the two-step process takes longer as more steps for production and purification are needed. Lipases are able to catalyse both esterification and interesterification reactions. Pancreatic PLA_2 functions reasonably well for esterification, but cannot catalyse interesterification/transesterification since no acyl–enzyme intermediate is formed (Adlercreutz *et al.*, 2003). However, it is possible to use this enzyme to exchange the fatty acid in the *sn*-2 position in a one-step reaction. Hydrolysis and re-esterification then occur in parallel.

To shift the equilibrium to the product side during acidolysis, product should be removed from the reaction system after its formation. However, the structured PLs are not easily separated from the system, as their properties are very similar to the substrate material. Alternatively, it may be possible to start with a high water content to facilitate formation of the intermediate lyso-PC. Subsequent removal of water from the system will facilitate the formation of diacyl-PLs.

2. *Process parameters for the modification of acyl groups*

Several important parameters are considered to influence enzymatic esterification and transesterification for the production of diacyl-PLs with modified fatty acid composition. These parameters include PL head group, acyl donor, enzyme dosage, solvents, water content, reaction temperature and reaction time. Reactions have so far been performed mainly in small-scale laboratory experiments, and little effort has been made to develop these reactions to a larger scale.

A literature evaluation of the enzymatic acyl modification of PLs has recently been undertaken (Adlercreutz *et al.*, 2003; Guo *et al.*, 2005). Several studies have reported relatively high yields under certain reaction conditions. Haraldsson and Thorarensen (1999) reported an incorporation of eicosapentaenoic acid of 53% and a yield of 63% in a solvent-free system by *Rhizomucor miehei* lipase-catalysed acidolysis with a substrate molar ratio of 6.7 (fatty acid/PC) and water content of 2% w/w. Similar incorporation and yield were reported by Svensson and co-workers (1992), using immobilized *Rhizopus oryzae* lipase in a solvent system (toluene) with a substrate molar ratio of 40 under controlled water activity (a_w = 0.06). For the esterification of LPC to form PC, even higher yields (~75%) have been reported with the same lipase when using an even higher substrate molar ratio (90) at higher water activity (a_w = 0.22) in a solvent system (Adlercreutz *et al.*, 2002). High yields (60%) have also been reported for esterification using immobilized PLA_2 in toluene with substrate ratio of 180 mol/mol oleic acid/LPC (Egger *et al.*, 1997). For the same enzyme, Hosokawa and co-workers (1995) reported a 40% yield of PC during esterification with glycerol as solvent and low substrate molar ratio (~3 mol/mol eicosapentaenoic acid/PC).

Increasing the fatty acid concentration increases the yield for both lipase-catalysed esterification and transesterification reactions, since hydrolysis is inhibited by this increase (Adlercreutz *et al.*, 2002). Reaction time for esterification reactions to reach 95% conversion was independent of the fatty acid concentration. However, during transesterification, reaction time increased with increasing fatty acid concentration. Decreased reaction rates may be caused by increased polarity of the reaction medium on addition of fatty acids. Decreased reaction rates have also been reported during the PLA_2-catalysed esterification reactions with increasing amounts of fatty acids, and were possibly caused by changes in polarity or viscosity (Egger *et al.*, 1997). In the solvent system, high substrate ratios did not seem to be a problem until very high ratios were employed, whereas inhibition is usually observed at lower substrate ratios for solvent-free systems (Peng *et al.*, 2002). The choice of substrate molar ratio also influences the downstream processing cost and difficulties in separating fatty acids (both acyl donor and exchanged fatty acids) from products. Therefore, a compromise usually has to be made for substrate molar ratio.

Water seems to have a complex role in terms of compromising enzyme activity, hydrolysis side reactions, reaction rate, and extent of incorporation (Haraldsson & Thorarensen, 1999). Lipases have been reported to operate even at very low water content, which inhibits hydrolysis. PLA_2 on the other hand requires a higher water activity to function and therefore the yield is expected to be lower (Adlercreutz *et al.*, 2003). From the engineering point of view, there will be an optimum water content in the system for optimum incorporation of fatty acids and yield of structured phospholipids. In some cases, it has been shown that with increased water content in the system there is increased incorporation of novel fatty acids, but with major hydrolysis in parallel. Two ways commonly used for controlling water activity are pre-equilibrium of reactants and enzyme with saturated salt solution or direct addition of salt hydrates with a definite a_w value (Adlercreutz *et al.*, 2002; Egger *et al.*, 1997). For scaled-up production, this would be impractical for the techniques so far developed. Others have simply added water directly to substrate or wetted the enzyme prior to addition to the reaction mixture (Peng *et al.*, 2002). Yamamoto and co-workers (2006) and Hosokawa and co-workers (1995) used water mimics, such as formamide, ethylene glycol or polypropylene glycol, for *in situ* control of water activity during esterification and acidolysis. These polar solvents activated the enzyme, but did not participate in hydrolysis, unlike water. This resulted in enhanced incorporation and recovery of diacyl-PLs. Recently, it was demonstrated that addition of certain proteins to the reaction medium may also increase the yield (Yamamoto *et al.*, 2006).

The use of solvent should be avoided when a solvent-free system can satisfy the reaction requirements. Solvent may increase reactivity through lowered viscosity and so increase mass transfer. An organic solvent environment allows better accessibility of enzymes to substrates and better availability of water at the active site of the enzyme. Presence of solvent improves mixing in the system, and makes the subsequent removal of enzyme more convenient. However, solvents with high polarity are not suitable for use in enzymatic esterification reactions, as they have a stronger water partitioning effect from the enzyme in the system and thereby inactivate the enzyme (Laane *et al.*, 1986). In general, activity decreases with increased polarity. Commonly used solvents for lipase and phospholipase-catalysed acidolysis and esterification are toluene and hexane, both having log $P > 2$ (Haraldsson & Thorarensen, 1999; Park *et al.*, 2000).

From a safety and health point of view, a solvent-free system is preferred, especially if the reaction is scaled up. In the solvent-free system, no organic solvent is present other than the substrate itself. During the acidolysis reaction, free fatty acids (FFA) are normally used in large excess (high molar substrate ratio), and function as solvent for the phospholipid. Compared to the solvent system, the reaction time may be longer since the amount of substrate is higher. In the solvent-free system, the PL concentration will be very high, probably causing major stripping of water from the enzyme and the reaction system, and may therefore require additional water (Haraldsson & Thorarensen, 1999).

Incorporation of novel fatty acids into phospholipids using a solvent-free system has been reported to be high during both lipase- and PLA_2-catalysed acidolysis reactions; however, the yield is considerably higher when using lipase as catalyst in comparison to using PLA_2 (Haraldsson & Thorarensen, 1999; Peng *et al.* 2002; Vikbjerg *et al.* 2007).

The viscosity of the substrate is dramatically reduced at higher temperatures, resulting in increased mass transfer and thereby increased reaction rate. Too high temperatures will however irreversibly denature the enzyme, resulting in reduced enzyme performance. It was reported that low temperatures were favoured to minimize the rate of hydrolysis and loss of product during interesterification of phospholipids (Vikbjerg *et al.*, 2005; Doig & Diks, 2003a). Enzymes in general exhibit dramatically enhanced thermostability in nearly anhydrous media (Mingarro *et al.*, 1994). This is explained on the basis of both increased conformational rigidity and minimization of deleterious covalent processes involved in irreversible inactivation.

3. Processes for the modification of acyl groups

The exchange of fatty acids in PLs has mainly been performed batch-wise in small screw cap vessels or glass bottles with orbital shaking or magnetic stirring. Operating batch-wise, the enzyme has to be separated from the medium containing the reaction product, normally by centrifugation or filtration. Product loss may occur in this step due to absorption on the enzyme carrier. The batch reactor has many advantages: it is simple to use and there is flexibility to change the operation from batch to batch. In this kind of reactor, the immobilized enzyme is dispersed in the reaction solution. The main disadvantage of batch reactors is that the reactor has to be emptied and refilled at the end of the batch run. This leads to down-time and a loss of productivity. In addition, batch reactors can suffer from batch-to-batch variations.

The packed-bed reactor (PBR) is the most frequently used reactor in large-scale operations for immobilized enzymes, because of high efficiency, low capital investment, ease of construction, continuous operation and easy maintenance. In a PBR, the immobilized enzyme particles are held in a column and substrate is pumped through under plug flow conditions. Heating is controlled with a water jacket or by inside heating elements on a larger scale. In a PBR with up-flow, the enzyme at the bottom of the column is exposed to substrate whereas the enzyme at the top is exposed largely to product. The main disadvantage of PBRs is the difficulty of maintaining even flow and constant temperature.

There are only a few reports on enzymatic modification of phospholipids

terminal phosphate diester bond of glycerophospholipids and releasing the polar head groups (Ulbrich-Hofmann *et al.*, 2005; Frohman & Morris, 1999). Depending on the acceptor (water/alcohol) of the phosphatidyl moiety, this reaction can lead to hydrolysis or transphosphatidylation (Figure 3.3). Transphosphatidylation allows the efficient and selective transfer of the phosphatidyl group to a compound with reactive hydroxyl groups, and accordingly provides a powerful tool for modifying the physicochemical properties of the compound. Therefore, from an application point of view, PLD-catalysed polar head group exchange of glycerophospholipids constitutes by far the most synthetically useful phospholipid transformation catalysed by phospholipases (Guo *et al.*, 2005; Servi, 1999).

Table 3.1 shows some representative alcohol structures for PLD-catalysed phosphatidyl transfers investigated so far. Phospholipase D has a very broad substrate spectrum from polar head moieties of natural PLs (e.g., choline, aminoethanol), through primary aliphatic alcohols, primary alcohols with other multiple functional groups, cyclic alcohols, heterocyclic or phenol structures, to sugar and nucleosides, even PL analogue alkylphosphate ester, and a variety of structurally diverse compounds can serve as the reactive substrates of PLD. This has greatly extended the application range of transphosphatidylation.

If the starting materials are natural PLs (e.g., soybean lecithin), the strategy for enhancing the yield of desired species from the mixture will become simple, with minimal use of toxic solvents using an enzymatic approach. Further purification can be omitted, because small amounts of impurities and unreacted materials from natural sources are generally regarded as safe and with acceptable biocompatibility (Servi, 1999). Enrichment of a particular PL species can be achieved by treatment of a PL mixture with the appropriate alcohol of the polar head group of the PL species in the presence of PLD (Doig & Diks, 2003b; Juneja *et al.,* 1992, Iwasaki *et al.*, 2003). Phosphatidylinositol is only hydrolysed by PI-specific PLD; PI can therefore be enriched by hydrolysis of other PL species by PLD with broad substrate specificities, leaving PI unchanged (Yamane *et al.*, 1990). The rate and selectivity of PLD-catalysed transphosphatidylation depends on the enzyme source, the nature and concentration of the alcohol acceptor, the properties of the reaction medium and other operational conditions. The main objectives of fundamental studies targeted on biochemical understanding are to explore the interdependence of enzyme activity, structural characteristics of acceptors, media and other important parameters. From the standpoint of application, the ultimate purpose is to construct a toolbox of techniques to maximize the yield and purity of desired products. Strategies to fulfil this aim have been progressively developed

Table 3.1. Representative structures of alcohol acceptors (HO-X) for enzymatic transphophatidylation [a]

Polar head moiety of natural PL species	Linear aliphatic alcohols	Primary alcohols attached in a functional structure	Cyclitols and phenols
1. 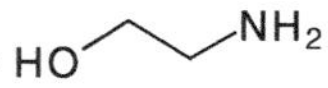	5. 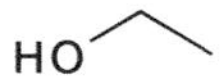	10. 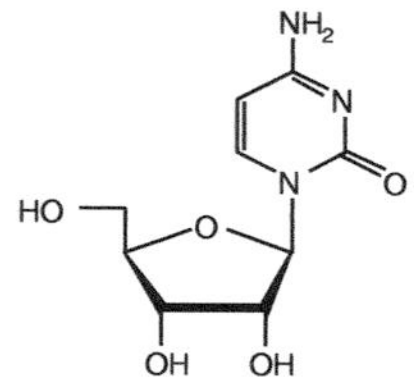	17. 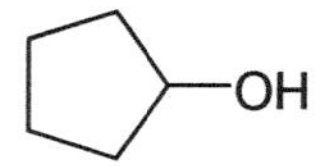
2. 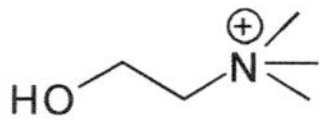	6. 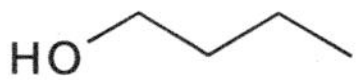	11. 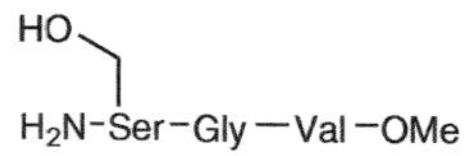	18. 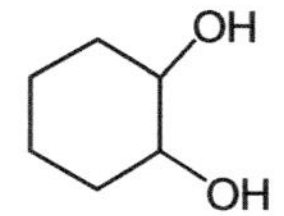
3. 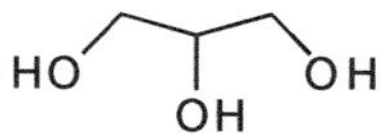	7. 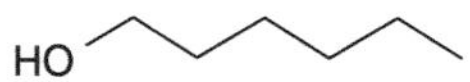	12. 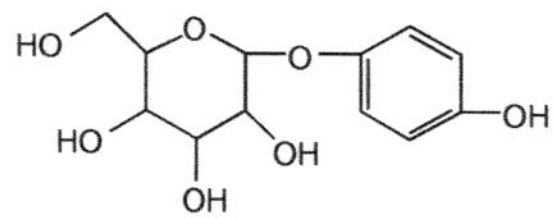	19. 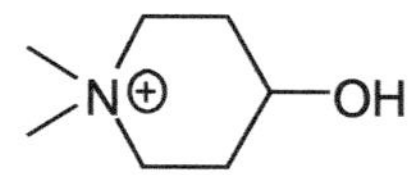
4. 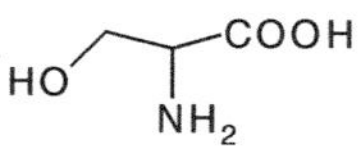	8. 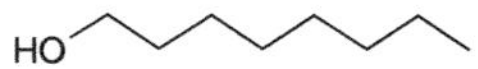	13. 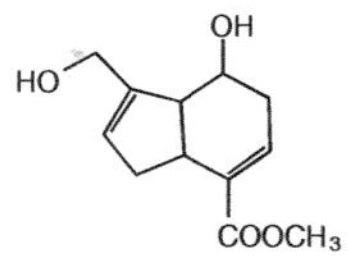	20. 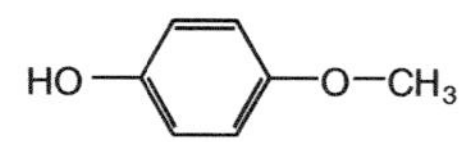
	9. 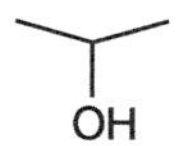	14. 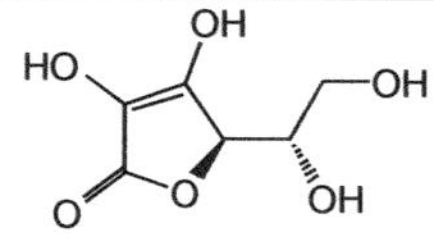	21. 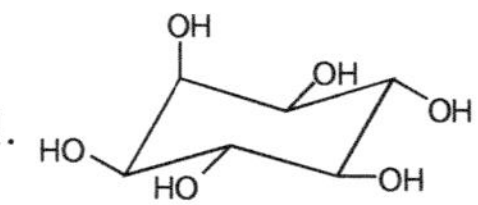
		15. 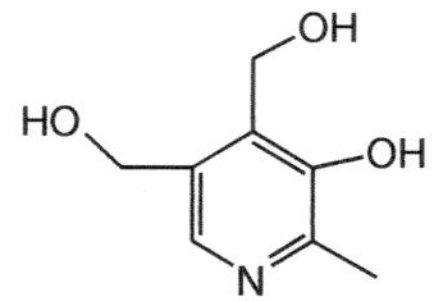	
		16. 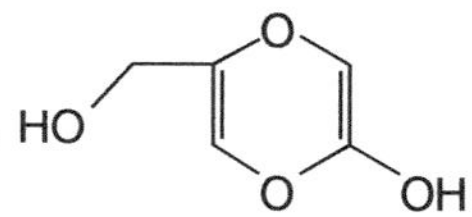	

[a] Compounds 1–21 are ethanolamine (1), choline (2), glycerol (3), serine (4), ethanol (5), butanol (6), hexanol (7), octanol (8), 2-propanol (9), ribosidocytosine (10), peptide of Ser-Gly-Val-OMe (11), arbutin (12), genipin (13), ascorbic acid (14), pyridoxine (15), kojic acid (16), cyclopentanol (17), 1,2-cyclohexanediol (18), 4-hydroxy-*N,N*-dimethylpiperidine (19), 4-methoxyphenol (20) and inositol (21), respectively. The corresponding references are listed in Table 3.2.

(D'Arrigo & Servi, 1997; Doig & Diks, 2003b; Guo *et al.*, 2005; Ulbrich-Hofmann *et al.*, 2005; Rich & Khmelnitsky, 2001). These will be summarized in the following section. The idea of preparing PLs bearing a pharmacologically active molecule in the polar head has produced a number of new molecules (Servi, 1999). The knowledge and experience acquired from these fundamental investigations and simple model reaction systems concerning enzyme species, acceptor structure and medium properties help develop strategies for implementing practical protocols of industrial interest (Hirche *et al.*, 1997; Hirche & Ulbrich-Hofmann, 2000; Dittrich & Ulbrich-Hofmann, 2001).

2. *Important parameters and processing technology developments*

(a) Phospholipase D source and specificity. Currently more than 500 PLD entries are found in the NCBI Genbank (Ulbrich-Hofmann *et al.*, 2005). Sequence comparison disclosed that PLD from different sources share highly close homologues and sequence motifs, (Ponting & Kerr, 1996). These enzymes contain two HxxKxxD motifs and it has been suggested that the conserved histidine, lysine and aspartate residues form a catalytic triad responsible for formation or hydrolysis of phosphoester bonds (Waite, 1999; Frohman & Morris, 1999). Based on the convenience of enzyme production, transphosphatidylation activity and storage stability, the PLD from plants and various strains of *Streptomyces* are the enzymes mainly used for synthetic purposes. Almost all PLD are Ca^{2+}-dependent and differ only in calcium concentration levels (Wang, 2000). PLD from plants require higher concentration of 20–100 mM Ca^{2+}, whereas the enzymes from mammals need concentrations of 1–9 mM Ca^{2+} and the PLD from yeast require little Ca^{2+}. Generally, the transphosphatidylation activities of PLD from *Streptomyces* species are considered superior to those of PLD from plant sources such as cabbage. However, it is dangerous to generalize, because enzyme sources, acceptor alcohols and reaction systems have proved to be strongly interdependent (Hirche & Ulbrich-Hofmann, 2000). For instance, the preference of PLD from *Streptomyces* spp. over cabbage PLD becomes small or diminishes in the exchange of choline in PC by secondary *N*-heterocyclic alcohols (Hirche & Ulbrich-Hofmann, 2000); in contrast, in the transesterification of alkylphosphate esters, PLD from cabbage and *Streptomyces chromofuscus* showed higher selectivity than PLD from *Streptomyces* spp. (Aurich *et al.*, 2002). With increasing recombination-expression and genetic engineering applied to PLD production, more isoenzymes from different sources have been expressed in *E. coli* (Zambonelli *et al.*, 2003; Schäffner *et al.*, 2002; Wang *et al.*, 1994). Different isoenzymes from the same source sometimes exhibit significant differences in their transphosphatidylation activity (Ulbrich-Hofmann *et al.*, 2005). The progress in molecular and structural biology of PLD, especially directed evolution, seems to offer the potential of engineered PLD with desired properties for transphosphatidylation (Ulbrich-Hofmann *et al.*, 2005).

(b) Substrate specificity and the structural characteristics of alcohol acceptors. Most PLD are capable of hydrolysing a broad range of PLs including PC, PE, PG, PI, PS, lyso-PC and cardiolipin (Waite, 1999).The preferred substrate is PC and there may be significant differences of the substrate specificities for the enzymes from different sources (Waite, 1999; Wang, 2000). Besides the natural substrates of 1,2-diacylglycerophospholipids, PLD can also act on ether PLs, such as plasmalogens (Waite, 1999), and even phosphate diesters and phosphonates (Dittrich *at al.*, 1998), which are not characterized by a glycerol backbone. However, these observations seem insufficient to conclude that the non-polar head part has little influence on substrate recognition of PLD, because PLD stereospecificity is completely lost when the 2-acyl group in the substrate is replaced by an alkyl group (Bugaut *et al.*, 1985).

The reactive hydroxyl groups in acceptors with different locations and configurations (Table 3.1) can be categorized into four groups, based on structural characteristics and applications. The first group, used for interconversion of natural PL species, is water-soluble primary alcohols. They are generally good substrates and can be easily incorporated into PLs with PLD, especially with PLD from *Streptomyces* sp. (Guo *et al.*, 2005). The transphatidylation activity of aliphatic alcohols with PC decreases significantly in two-phase systems with chain lengthening (Okahata & Mori, 1998). The third group of alcohols with functional attachments represent the most interesting and useful alcohol acceptors (Wang *et al.*, 1993; Servi, 1999). These compounds are either pharmaceuticals or biologically active molecules. The steric hindrance of the primary hydroxyl group in the compound principally governs the reaction rate (D'Arrigo *et al.*, 1994). If there are multiple primary hydroxyl groups in a substrate, the reaction favours the transfer of phosphatidyl to the hydroxyl group with minimum steric hindrance (Servi, 1999). It should be noted that all secondary alcohols successfully used for PLD-catalysed transphosphatidylation are cyclitols, heterocyclic alcohols or similar structures (D'Arrigo *et al.*, 1994; Rich & Khmelnitsky, 2001; Okahata & Mori, 1998). Among secondary linear aliphatic alcohols, only alkan-2-ols (containing the unit $CH_3CH(OH)$-) were capable of being incorporated at different yields (D'Arrigo *et al.*, 1994; Rich & Khmelnitsky, 2001; Okahata & Mori, 1998). Okahata and Mori (1998) compared the structures of aliphatic and cyclo-alcohols on transphosphatidylation rate, revealing that the reaction rate decreases from cyclobutanol to cyclohexanol. Cyclobutanol achieves a comparable rate with myristyl alcohol and cyclopentanol is comparable with palmityl alcohol, indicating a profound effect from different steric hindrances of different alcohol structures. Takami and Suzuki (1995) successfully prepared 1,2-dipalmitoylphosphatidyl-4-methoxyphenol at 45% yield using PLD from *Streptomyces* sp. as catalyst, suggesting phenol also can be a reactive acceptor. Although the preparation of PI by PLD-catalysed transphosphatidylation using inositol is of great interest, reports on this topic are contradictory (D'Arrigo *et al.*, 1994). Attempts to prepare phosphatidylinositols using PLD from *Streptomyces* sp. failed, but PLD isolated from spinach leaves was reported to give PI from PC and inositol (Mandal *et al.*, 1980).

(c) Effects of reaction media. PLD-catalysed transphosphatidylations are mostly performed in two-phase emulsion systems, where organic solvent containing the dissolved PLs is mixed with the aqueous phase containing the enzyme and the acceptor alcohol in the buffer solution. A lot of previous work has shown that solvents decisively affect the course of the reaction (Servi, 1999; Ulbrich-Hofmann *et al.*, 2005). Ulbrich-Hofmann and co-workers examined the correlation of the interfacial pressure of different solvents and the package density of the substrate aggregates and found that the different arrangements and packing densities of PL aggregates had a profound role in regulating enzyme activity (Hirche & Ulbrich-Hofmann, 1999; Hirche *et al.*, 1997). Their work also showed that the physicochemical properties of the organic solvents decisively influenced the ratio of transphosphatidylation to hydrolysis rate. For amphiphilic acceptor alcohols such as butanol, transphosphatidylation is faster than hydrolysis in a non-polar solvent such as hexane. For water-miscible acceptor alcohols such as glycerol, transphosphatidylation is faster than hydrolysis in more polar solvents such as diethyl ether. For secondary alcohols (e.g., secondary *N*-heterocyclic alcohols), hydrolysis is stronger for *Streptomyces* sp., but for PLD from cabbage, on the contrary, hydrolysis is less strong.

(d) Process technology development. Due to its high PLD preference and its commercial availability, PC is the most common starting material for transphosphatidylation. However, choline, produced as a by-product, easily accumulates in the immediate vicinity of the enzyme. This inhibits the activity of the enzyme and results in lower conversion (Rich & Khmelnitsky, 2001). Continuous removal of choline from the reaction mixture will drive the reaction to the product side and complete conversion can be expected. For this reason, choline oxidase and catalase have been used in the process. Oxidase can oxidize choline first into betainealdehyde and then to betaine, liberating hydrogen peroxide. Catalase can degrade hydrogen peroxide to water and oxygen. Better performance has been obtained (Juneja *et al.*, 1992). Another interesting protocol, which originated from the suggestion of employing ion exchange resin as PE scavenger (Shigematsu & Hasegawa, 1989), is the use of a cation-exchange resin serving as a scavenger for choline (Rich & Khmelnitsky, 2001). The reaction thus could be performed in anhydrous organic solvent. The undesirable hydrolysis was avoided and high yields were obtained. Another patented process to obtain highly pure PG from PC is to prepare a phosphatide calcium precipitate first and then to convert this precipitate to its corresponding soluble salt by ion exchange in the organic solvent (Tremblay *et al.*, 1990).

To avoid the uses of solvents in reaction systems, a pure aqueous suspension system has been developed with high reaction efficiency (Iwasaki *et al.*, 2003). Calcium sulfate was used to form the aqueous suspension system. With this system up to 180mg/ml lecithin can be completely converted to PS in 24 h. The modified PS can be easily extracted from the powder. A typical system for reaction includes: 7.2 mg/ml of lecithin, 3.87 M serine, 50 mM acetic-sodium acetate buffer (pH 5.6), 36 mg/ml calcium sulfate powder, and 1.5 units/ml PLD. The mixture is reacted at 40°C with stirring for 24 h.

Due to the high price of PLD and its unavailability in an immobilized form, approaches using PLD as a biocatalyst aim at products with high additional value, especially for pharmaceutical applications. Although some alternative protocols have been suggested to avoid using toxic solvents, they usually need extra procedures. Therefore, apart from greatly reducing the cost of PLD, simplifying the performing steps and increasing the recovery and reusability of enzyme are still a problem.

2. *Recent progress for the polar group modification*

Table 3.2 summarizes representative synthesis or modification studies for different purposes. These studies covered attempts to enrich a particular PL species, to explore the interdependence of enzymes, solvents and alcohol acceptors, and to produce novel PL-containing biologically active compounds or pharmaceuticals, or alkylphosphate esters for medical applications. This summary provides markers to track progress aiming at the exploration of innovative applications and the development of efficient reaction systems of PLD-catalysed transphosphatidylation.

E. Enzymatic modification of sphingolipids

A great deal of physiological research has been conducted on sphingolipids because of their many biological functions (Chapter 8). The results from this

Table 3.2. Literature evaluation for PLD-catalysed transphosphatidylation

Acceptor alcohols	PLD	Media	Yield/conversion	Remarks	Reference
Choline	SA-PLD (Honen)	Ethyl acetate/water buffer biphasic system	100% for PE as substrate and 0% for PI as substrate	For increasing PC content in commercial lecithin	Juneja *et al.*, 1989
Serine	SA-PLD (Honen)	Ethyl acetate or diethyl ether/water buffer biphasic system	80% (30 min)	PS has various functions for brain health	Juneja *et al.*, 1992
Serine, ethanolamine, glycerol	SS-PLD (Sigma)	Hexane/1-octanol or ethyl acetate/water buffer biphasic system	73% (PG), 72% (PS), 77% (PE)	Toolbox build-up for modification of lecithin headgroups	Doig and Diks, 2003b
Serine	SS-PLD (Asahi)	Ethyl acetate/ water buffer biphasic system	40% (6 h)	PS with DHA on the *sn*-2 position having a higher acetylcholine-releasing effect for mouse brain synaptosomes	Hosokawa *et al.*, 2000
Serine	SS-PLD (Sigma)	Water buffer system	Not listed	PLD immobilized by cross-linking	Yaqoob *et al.*, 2001

Serine	SA-PLD (lab-made)	Water buffer system with $CaSO_4$ suspension	80% (24 h)	Product being an activator of protein kinase C ande regulating the activity of various enzymes and playing an important role in mental functions in the brain	Iwasaki *et al.*, 2003
Serine, glycerol, ethanolamine	SS-PLD, PLD from peanut and cabbage (Sigma), and SA-PLD (lab-produced)	Chloroform, diethyl ether, or ethyl acetate/water buffer biphasic system	99% (1 h)	For the chemoenzymatic synthesis of PLs bearing acetylenic fatty acids in the stabilization of membrane proteins	Pisch *et al.*, 1997
Glycerol	SS-PLD (Lipoid)	Water buffer system	60–72%	PLD immobilized by covalent binding to VA-Epoxy Biosynth or porous glass	Dittrich and Ulbrich-Hofmann, 2001
Glycerol	Savoy cabbage (self-extracted)	Water buffer system	Low	Product with improved emulsification properties	Virto and Adlercreutz, 2000
Primary, secondary and aromatic alcohols	SS-PLD (Asahi)	Benzene or chloroform or hexane/water buffer biphasic system	90% (1 h) for 1-butanol (similar for up to 10 carbon aliphatic primary alcohols)	Lipid-coated PLD to improve the solubility and transfer in solvent phase	Okahata *et al.*, 1995
Primary, secondary and aromatic alcohols	SS-PLD (Asahi)	Anhydrous chloroform system	52–100% (1–2 days)	Uses of cation-exchange resin as a scavenger for choline	Rich and Khmelnitsky, 2001

Table 3.2. *continued*

Acceptor alcohols	PLD	Media	Yield/conversion	Remarks	Reference
Secondary alcohols	SS-PLD (Type VII, Sigma) and SS-PLD (Asahi)	Ethyl acetate or dichloromethane/ acetate buffer system	52–95% depending on alcohol structural specificity	β-Alipahtic alcohols, cyclic mono- and di-alcohols are reactive acceptors. Reactivity of γ-aliphatic alcohols, inosintol and cyclic multialcohols were very low. The water solubility, steric hindrance and nucleophilicity of acceptors were suggested to be responsible for transformation efficiency.	D'Arrigo *et al.*, 1994
Dihydroxyacetone	SS-PLD (Asahi)	Ethyl acetate/ water buffer biphasic system	70% (2 h)	Acceptor alcohol being a tanning agent in cosmetic preparations and an anti-oxidant and a free radical scavenger	Takami and Suzuki, 1994a
1-Lauroyl-*rac*-glycerol, 1-lauroyl-dihydroxy-acetone	SS-PLD (Sigma)	Diethyl ether/water buffer biphasic system	90% (4 h) and 80% (20 h), respectively	For the production 1-lauroyl-*rac*-glycero-phosphate and 1-lauroyl-dihydroxyacetonphosphate	Virto and Adlercreutz, 2000

4-Methoxyphenol	SS-PLD (Asahi)	Benzene one phase system	45%	PLD immobilized on Amberlite IRC-50, a cation-exchange resin. Test on water-insoluble acceptor alcohols	Takami and Suzuki, 1995
N-Heterocyclic alcohols or As-containing group	SS-PLD (Type VII, Sigma)	Diethyl ether/water buffer biphasic system	Up to 80% for most acceptors studied	Acceptor alcohols with antitumour activity	Hirche *et al.*, 1997
Genipin	SS-PLD (Asahi)	Ethyl acetate/water buffer biphasic system	80% (1 h)	Product having 6–52 times stronger cytotoxity than genipin to Hela cells, human embryonic fibroblasts, and MT-4 cells	Takami and Suzuki, 1994b
Kojic acid and arbutin	SS-PLD (Asahi)	Ethyl acetate (arbutin) or diethyl ether (kojic acid)/ water buffer biphasic system	80% at 1 h for kojic acid with CH_2Cl_2 as solvent	Both acceptor alcohols having inhibitory activity to tyrosinase	Takami *et al.*, 1994
Ascorbic acid	SL-PLD (Honen), SS-PLD (Toyo Jozo, Sigma)	Diethyl ether/water buffer biphasic system	80% (6 h)	Bioactive antioxidants for membrane lipid peroxidation, important antioxidant in the cellular defence against active oxygen radicals	Nagao *et al.*, 1991

Table 3.2. *continued*

Acceptor alcohols	PLD	Media	Yield/conversion	Remarks	Reference
Azasugars, nucleosides, and peptides	SS-PLD (Sigma)	Chloroform/water buffer biphasic system	Various from 15% to 98% related to types of the acceptor alcohols	Synthesis of PL-inhibitor conjugates targeting on liposome applications, azasugars being inhibitors of glycosidase enzymes and having potential utility in the treatment of metabolic disorders	Wang *et al.*, 1993
1,8-Octanediol, then condensed with sialic acid-based structure	SS-PLD (Honen)	Chloroform/acetate buffer system	67.8% for trans-phosphatidylation	Chemical and enzymatic synthesis of antirotaviral agent: sialylphospholipid	Koteksu *et al.*, 1997
Saccharides	AS-PLD (Sigma)	Diethyl ether/water buffer biphasic system	85% for glucose with PC as substrate (20 h)	Pentoses, hexoses, and heptoses with primary alcohol groups being good acceptor alcohols, PE, PG, and SM having approximately ⅔, ⅓, and ⅕ reactivity of PC, respectively	Kokusho *et al.*, 1993
N-heterocyclic alcohols	PC-PLD (Sigma)	Hexane/1-octanol/water buffer biphasic system	82% (24 h)	For the production of alkylphosphate esters	Aurich *et al.*, 1997

Abbreviations: AS-PLD, *Actinomadura* sp.; SA-PLD, *Streptomyces antibioticus*; SS-PLD, *Streptomyces* sp.; PC-PLD, *Streptomyces chromofuscus*; SL-PLD, *Actinomadura lydicus*. PLD, phospholipase D; PS, PC, PE, PG, phosphatidyl serine, choline, ethanolamine and glycerol, respectively; SM, sphingomyelin; PL, phospholipid; DHA, docosahexaenoic acid.

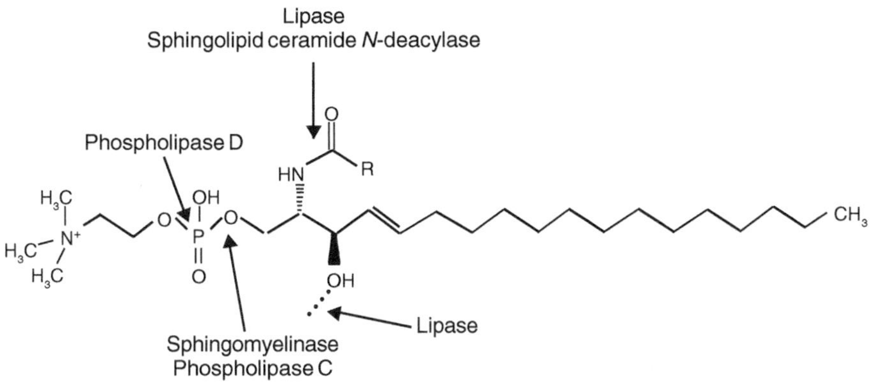

Figure 3.4. Enzymes used for the modifications of sphingomyelins and corresponding modification sites. Dotted line indicates the possible esterification reactions.

research have led to the development of industrial applications of sphingolipids and their derivatives. Furthermore, many biologically active sphingolipids and relevant enzymes are now commercially available. Therefore, the situation for the technical and engineering study of sphingolipid modifications is developing rapidly. As shown in Figure 3.4, several enzymes have different functions in sphingolipid modifications, and some examples of their modifications will be introduced below.

Lysosphingolipids, which themselves have many biological functions (Heringdorf *et al.*, 2002), can be used as starting materials for synthesizing novel sphingolipids (Nakagawa *et al.*, 2005). Sphingolipid ceramide *N*-deacylase produced from *Pseudomonas* sp. TK4 can efficiently cleave the *N*-acyl linkage of ceramides in various glycosphingolipids and sphingomyelins to yield their lyso forms (free amines) and fatty acids (Ito *et al.*, 1995) (Figure 3.4). This enzyme is different from ceramidase, which has catalytic activity only towards free ceramide but not towards other sphingolipids (Hassler & Bell, 1993). Moreover, sphingolipid ceramide *N*-deacylase has also been found to catalyse the acylation of lysosphingolipids, which is the reverse reaction of the hydrolysis (Kita *et al.*, 1996). The enzyme has been applied to the production of lysosphingolipids and the synthesis of novel ceramides and sphingolipid derivatives (Mitsutake *et al.*, 1997; Ito *et al.*, 1999). With the same enzyme, sphingolipid hydrolysis proceeds more efficiently at acidic pH in a high detergent concentration, whereas the preferred conditions for the reverse reaction are at neutral pH with a low detergent concentration (Kita *et al.*, 2001). Therefore, the direction of the catalytic reaction can be controlled through the manipulation of reaction conditions and substrate concentrations. Consequently, the production of novel sphingolipids can be achieved in two steps – the hydrolysis of natural sphingolipids and the acylation of the reaction intermediate (lysosphingolipids) – using the same enzyme in different conditions. Ito and co-workers (1999) also proposed the direct transacylation between two sphingolipids or between one sphingolipid and one fatty acid, where the catalyst used was sphingolipid ceramide *N*-deacylase.

Due to their major role in maintaining the water-retaining properties of the epidermis (Coderch *et al.*, 2003), ceramides are of considerable commercial potential in cosmetics and pharmaceuticals. They have been broadly used as a moisture-retaining ingredient for human skincare products. Labelled ceramide has been synthesized from sphingosine and labelled fatty acid using sphingolipid ceramide *N*-deacylase produced from *Pseudomonas* sp. TK4 (Mitsutake *et al.*, 1997). The product can be used to detect ceramidase activity in biological samples. Ceramides and hybrid ceramides can also be obtained by the selective acylation of the amino group of lysosphingolipids using a commercially available lipase (Smeets *et al.*, 1997). The products have the proper stereochemistry, and therefore more closely resemble the natural compound. These synthetic methods need sphingosines or their derivatives as the starting material to produce ceramides.

Syntheses of isomerically pure sphingosines and their derivatives are tedious and time-consuming for industrial application (Elimelech, 2002). This limits the reduction of cost and further improvement of the ceramide production. An alternative method for the production of ceramides has been developed through enzymatic hydrolysis of sphingomyelins (Hidehiko *et al.*, 1995; Susumu *et al.*, 1999; Zhang *et al.*, 2006). Sphingomyelins are a ubiquitous membrane lipid and present in dairy products or by-products. Enzymes, such as phospholipase C and sphingomyelinase (EC 3.1.4.12), can break the bond between the primary hydroxyl group of ceramides and choline phosphate esters in sphingomyelins to generate ceramides (Figure 3.4). The hydrolysis of sphingomyelins has been improved through evaluation and optimization of several important factors, and phospholipase C from *Clostridium perfringens* shows higher catalytic activity towards the hydrolysis reaction than a few PLC enzymes from other sources (Zhang *et al.*, 2006). The reaction is more efficient in two-phase (water/organic solvent) systems than in one-phase (water-saturated organic solvent) systems. The reusability of the enzyme has been enhanced through immobilization on a carrier. Using Lewatit VP OC 1600 (Lanxess AG, previously Bayer AG, Germany) as the carrier, the immobilized enzyme retains around 70% of its initial activity after seven cycles (Zhang *et al.*, 2007).

Sphingosine-1-phosphate and its derivatives are valuable materials in the preparation of pharmaceutical compounds (Michaelis *et al.*, 2004) and cosmetic compositions (Kim *et al.*, 2003). Sphingosine-1-phosphate has been prepared from sphingosine in two steps involving the conversion of sphingosine to phosphatidylsphingosine catalysed by PLD and enzymatic hydrolysis of the intermediate to a diacylglycerol and sphingosine-1-phosphate with PLC (Morigaki *et al.*, 1999). Since sphingosine kinase (EC 2.7.1.91) is now available commercially (BPS Bioscience Inc. USA), sphingosine-1-phosphate can be easily obtained through the phosphorylation of sphingosine catalysed by sphingosine kinase. After phosphorylation, the product, sphingosine-1-phosphate, remains in the aqueous phase after solvent extraction (78% extraction efficiency), with very little contamination from sphingosine (Vessey *et al.*, 2005).

Phosphoryl-L-serine-*N*-acyl-sphingosine can be used in the trestment of many diseases of the central nervous system, such as Huntington, Alzheimer's and Parkinson's syndromes (Romeo *et al.*, 1994). To produce this valuable compound, PLD has been used to catalyse the modification of the polar group of sphingomyelins, where the choline moiety in sphingomyelins was replaced by serine in ethyl acetate buffer mixture at pH 5.6 and the presence of Ca^{2+} (Romeo *et al.*, 1994).

F. Enzymatic modification of glycolipids

In nature, there are many types of saccharides attached to a lipid moiety. Glycolipids of different forms are mostly abundant in plant cells and some species of bacteria. Plant glycolipids mainly consist of monogalactosylmonoacylglycerol (MGMG), monogalactosyldiacylglycerol (MGDG), digalactosylmonoacylglycerol (DGMG), digalactosyldiacylglycerol (DGDG) and sulfoquinovosyldiacylglycerol, in which MGDG and DGDG are major components. In contrast to plant glycolipids which are also termed as galactolipids, bacterial glycolipids contain mainly glucose instead of galactose, while the lipid part is more or less the same. The structures of the two major components are presented in Figure 3.5.

Most studies so far have been on types, structures, metabolism and cell functions of different glycolipids present in plants, bacteria and fungi. One of the best sources of information on such issues is a book by Kates (1990). In the past decade, many more studies have been reported on the discovery of novel and interesting properties of glycolipids of plant, marine and bacterial origins. These novel properties cover anti-cancer, anti-tumour, and anti-inflammatory bioactivities, modulation of immunological response, and biological antioxidation/heat protecting (Colombo *et al.*, 1996, 1998a, 1998b, 2001, 2002; Matsufuji *et al.*, 2000; Ramm *et al.*, 2004; Weil *et al.*, 2005). Some of

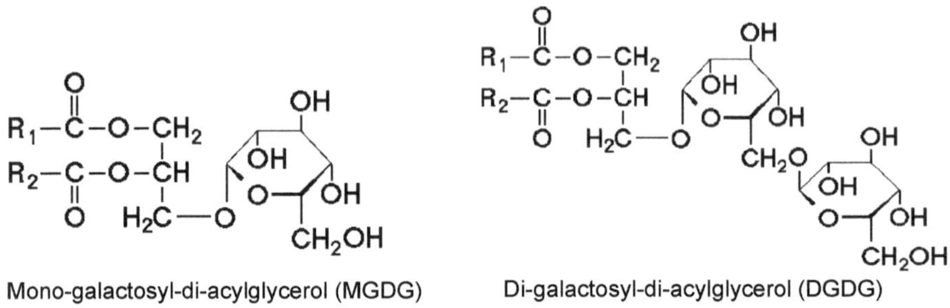

Figure 3.5. Chemical structures of galactolipids (DGDG and MGDG).

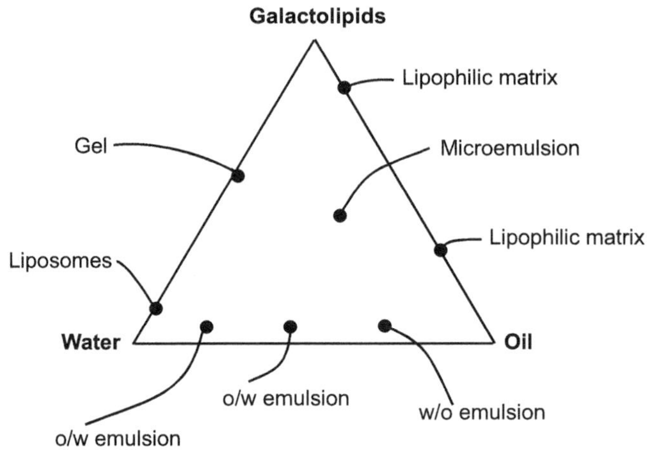

Figure 3.6. Physical properties and phase behaviour of plant galactolipid mixture (mainly DGDG and MGDG) (adapted from http://www.lipid.se/).

them, such as glycosyldiacylglycerol from rose hips, have been patented for their use as drugs in pharmaceutical preparations for arthritis (Larsen *et al.*, 2006). Furthermore, some of the plant galactolipids possess excellent physical properties for forming bilayers, hydrophilic lipid nanotubes, liposomes, hydrophilic and hydrophobic gels and dispersions, which are of much importance in pharmaceutical, food and cosmetic applications. Figure 3.6 shows the biphasic properties of the plant galactolipid mixtures (Carlsson *et al.*, 1997, 2000; Carlsson & Herslof, 1998; Fujima *et al.*, 2006).

The biological activities of galactolipids and glucolipids depend on types of both their fatty acids and saccharide components. There have been extensive studies on how the bioactivity of glycolipids is affected by changes in the type and number of saccharide moieties, stereomeric configurations of sugaracylglycerol, regioisomeric distribution of fatty acids in glycoglycerolipids, chain lengths and structures of fatty acids present in glycoglycerolipids (Colombo *et al.*, 1996, 1998a, 1998b, 2001, 2002, 2006; Matsufuji *et al.*, 2000; Ramm *et al.*, 2004). These studies form the rational basis for the interest in modification of these native glycoglycerolipids or synthesis of novel ones. To enhance or customize their desirable functionality, structural or compositional changes are required.

1. Modification (acylation, deacylation and fatty acid exchange) of native glycoglycerolipids

Enzymatic hydrolysis of native glycoglycerolipids of marine and plant origins by using fungal and bacterial lipases have been reported in recent years. Ramm and co-workers (2004) reported the modification of marine glucoglycerolipid using immobilized *Candida antarctica* lipase (Novozym 435) in a two-stage process. The entire reaction was conducted in *tert*-amyl alcohol at 55°C with less than 1% water; a yield of 84% deacylation product (diglycosylglycerol) was achieved. After purification of the deacylated product, reacylation was carried out to incorporate the new fatty acid (4-pentenoic acid) using the same lipase but with water content of lower than 0.1% together with molecular sieves in the reaction system. A yield of 88% reacylated product was obtained from the reaction under conditions at 50°C for 54 h. It should be noted that the reacylated products contained fatty acid acylated both at the *sn*-1 position of glycerol and the 6' position of the outer glucose moiety. The extent of acylation was observed to be dependent on the reaction time. Triacylated products also formed in various yields with various isomers after a very long reaction time.

In a similar study, the differences in specificity of different lipases were demonstrated using 2-*O*-beta-D-galactosyl-*sn*-glycerol (Colombo *et al.*, 2006). *Pseudomonas cepacia* lipase had less specificity in the acylation reaction illustrated in Figure 3.7 with conversion of 87% reactants, yielding 38% monoesters (*sn*-1-monoester and 6' -monoester in a ratio of 68:30) and 47% *sn*-1 and 6'-diesters, whereas in a similar reaction using Novozym 435 as biocatalyst, 85% conversion of reactants was achieved, with 67% reaction product as monoester containing more than 80% *sn*-3-monoester, thus showing high regiospecificity of the acylation reaction.

In order to achieve high regiospecificity of glyceryl–glycoside acylation, Morimoto and co-workers (1995) proposed the use of vinyl palmitate as acylating agent instead of conventional palmitic acid. They also screened five different lipases including *Achromobacter* sp., *Alcaligenes* sp., *Pseudomonas cepacia*, *Mucor javanicus* and *Rhizopus delemer*. Among these, *Acromobacter*

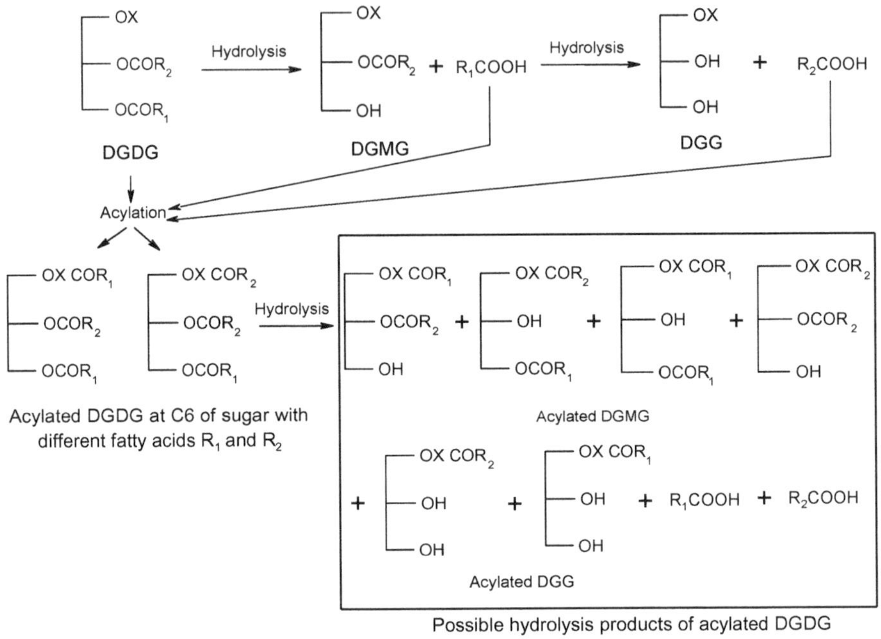

Figure 3.8. Possible hydrolysis and interesterification reactions of DGDG using lipases. X = digalactose moiety; R_1 and R_2 = fatty acids; DGDG = digalactosyldiacylglycerol; DGMG = digalactosylmonoacylglycerol; DGG = digalactosylglycerol.

sugars such as 2-*O*-acetyl-3,4,6-tri-*O*-benzylglucopyranosyl (Colombo *et al.*, 1995; Lafont *et al.*, 2006; Marino-Albernas *et al.*, 1996). It is worth mentioning that these protected sugars were used in the glycosylation of the glycerol/glycidol moiety with the carbohydrate aiming at glycosylation at the 1' position. However, this protection also helps in preventing fatty acid attachment to

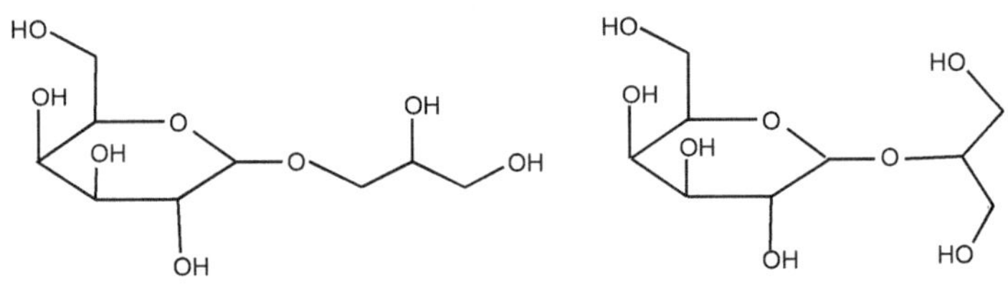

Figure 3.9. Isomers of galactosyl-glycerol.

the remaining hydroxyl groups of the carbohydrate during the fatty acid acylation step. On the other hand, the use of phenylboronic acid has been reported to be effective in preventing acylation at the 6' position of the carbohydrate in lipase-catalysed acylation reactions (Vaz *et al.*, 2000). In the study, immobilized Lipase-80000 was used as catalyst and the addition of phenylboronic acid in equimolar concentration to DGDG reduced acyl migration at the 6' position. There was a marked decrease in acyl migration from 35% to 5% in the production of 6' -acylated-DGDG and improvement in the fatty acid exchange at the *sn*-1 position of DGDG. The mechanism of this prevention by use of phenylboronic acid is not fully known, but complex formation between carbohydrate and phenylboronic acid might be responsible (Adlercreutz *et al.*, 2001).

3. De novo *synthesis of glycoglycerolipids using enzymes and future prospects*

The unmatched bioactive and excellent biphasic properties of glycoglycerolipids suggest that they have potential as food and pharmaceutical ingredients. Attempts to extract and purify glycoglycerolipids from plant and marine sources have not been very successful. The extraction and purification steps of these polar lipids are tedious and often expensive with low yield, thus limiting their industrial exploitation. Chemical and chemo-enzymatic synthesis of glycoglycerolipids at laboratory level has been well documented. (Hronowski *et al.*, 1991; Paltauf & Hermetter, 1994; Morimoto *et al.*, 1995; Marino-Albernas *et al.*, 1996; Lindberg *et al.*, 2002; Li *et al.*, 2003; Dubber *et al.*, 2006; Lafont *et al.*, 2006; Wandzik *et al.*, 2006). Studies on enzymatic synthesis of glycoglycerolipids with *de novo* procedures have not yet been reported. In the quest for an industrially viable and eco-friendly process for the production of these molecules, enzyme technology seems promising.

In nature, these glycolipids are synthesized by glycosyltransferases (GTases) catalysed reactions *in vivo* by monoglycosyldiacylglycerol transferase and diglycosyldiacylglycerol transferase, for example. Many of these GTases have multiple roles in the synthesis of various bioactive compounds including oligosaccharides, glycolipids and even, in some cases, glycopeptides. These glycosyltransferases are difficult to work with *in vitro* since they utilize nucleotide sugar (UDP-glucose and UDP-galactose) instead of a simple sugar as substrate, rendering their limited industrial exploitation. In spite of this fact, there have been attempts to use these GTases in the synthesis of different glycosides *in vitro* using various non-natural (non-nucleotide sugars) substrates. (Nakajima *et al.*, 1997; Thiem & Gambert, 1997; Jorasch *et al.*, 2000; Hu &

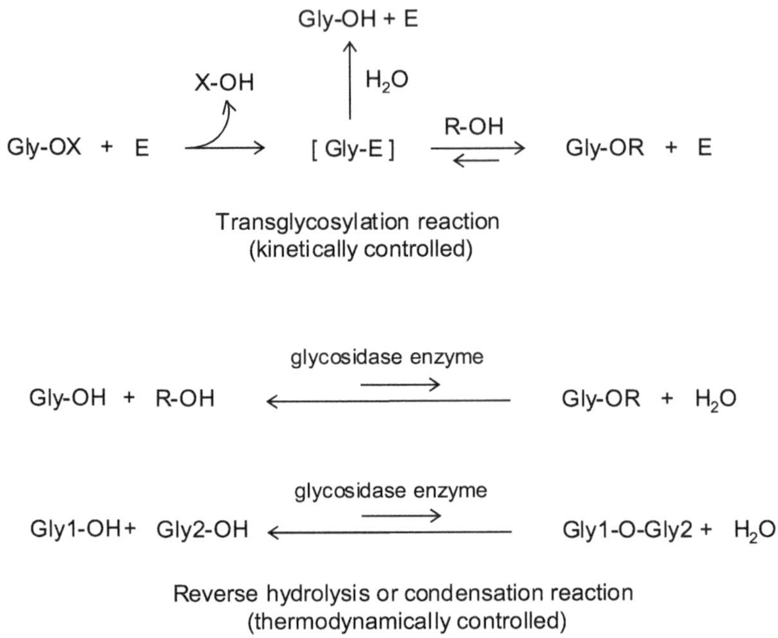

Figure 3.10. Depiction of transglycosylation and reverse hydrolysis reactions for synthesis of different glycosides. E = enzyme; R-OH = aglycon; Gly-OX, Gly1-OH and Gly2-OH = glycosyl donors; [Gly-E] = glycosyl enzyme complex; Gly-OR and Gly1-O-Gly2 = desired glycoside.

Walker, 2002; Qian *at al.*, 2002; Benning & Ohta, 2005; Pisvejcova *et al.*, 2006). In future, we may see these enzymes being engineered to work with easily available commercial sugar substrates. However, there are other classes of enzymes with the potential for the synthesis of glycoglycerolipids. In the past few decades, enzymatic transglycosylation and reverse hydrolysis (condensation) have been widely studied. Some transglycosidases (TGases) are industrially used for production of oligosaccharides. Another approach is to use the reverse hydrolysing potential of glycosyl hydrolysing enzymes such as α and β-gluco- and galactosidases. This reverse hydrolysing reaction takes place when water is replaced with another acceptor nucleophile having a free hydroxyl (-OH) group, such as mono- and disaccharides or alcohols. Figure 3.10 shows the differences between transglycosylation and reverse hydrolysis or enzymatic condensation. So far, major efforts have been made to synthesize and develop various glycosides and oligosaccharides. Detailed information on synthesis strategies can be found in the literature (Koeller & Wong, 2000; de Roode *et al.*, 2003; Garcia-Junceda *et al.*, 2004; Rowan & Hamilton, 2006).

As already discussed, the lipase-catalysed esterification of glycosylglycerol is possible. Thus, if the enzymatic synthesis of glycosylglycerol using a suitable transglycosylation or reverse hydrolysis reaction is combined with the lipase-catalysed esterification (acylation) of the formed glycosylglycerol, a glycoglycerolipid can be formed. There have been studies on glycoside synthesis between mono-disaccharide and glycerol (and other alcohols as well). The yields of the desired final product and flexibility of the reaction is complicated because both transglycosylation and reverse hydrolysis reaction are not easy to handle efficiently. Glycidol (2,3-epoxy-1-propanol) has been preferred over glycerol because it yields pure 3-*O*-glycoside due to the self protective effect of the epoxy group at the *sn*-1 and *sn*-2 positions. In addition, the use of glycidol makes the separation step easier due to its chloroform solubility (Silhavy & Boos, 1973; Fogarty & Benson, 1982; Bjorkling & Godtfredsen, 1988; Takenaka & Uchimaya, 2000; de Roode *et al.*, 2001; Nakano *et al.*, 2003; Yasutake *et al.*, 2003).

Direct attachment of the saccharide moiety with mono- or diacylglycerols is a difficult task, since it is hard to combine both reagents in a single soluble mixture for synthesis. Efforts have been made in a highly concentrated (also called plasticized glass phase) system where ethyl or propyl galactoside was used to couple with monoacylglycerol (Gill & Valivety, 2000). This area remains interesting and much unexplored in terms of finding a suitable technique for direct synthesis of glycoglycerolipid from a mono- or disaccharide and mono- or diacylglycerols. More work is definitely needed to explore the potential techniques.

G. Remarks and future perspectives

The enzymatic modification of PLs and other related polar lipids has attracted considerable interest for quite some time. The industrial implementation of enzyme technology for process and product development has also expanded quickly. However, technological knowledge is not always publicly available. The processes and technology for practical uses are mainly kept in industry. In academic studies, a great deal of information for product development is available, but technical evaluation of such information remains challenging. This is an area in which both biochemists and chemical engineers should increase their efforts. Two recent relevant reviews are recommended (Guo *et al.*, 2005; Song *et al.*, 2005).

Progress in lipid biochemistry clearly has not only disclosed the biological importance of PL and other related polar lipids for life and their contribution to human health, but has also provided many new opportunities for the development of polar lipid enzymology and engineering. These opportunities include the synthesis of lipid analogues of naturally occurring bioactive compounds for potentially pharmaceutical applications, and the provision of powerful biocatalysis tools for enzymatic applications through evaluation of novel enzyme sources and characterization of phospholipases and other related enzymes. However, it should be acknowledged that large-scale applications of biocatalysis in polar lipid modification are far from completion. The key issue for industrial application of enzymatic approaches is the high cost of enzymes and the development of improved enzymes. It is exciting that protein engineering and directed evolution may offer a solution to the problem (Song *et al.*, 2005). A large amount of work has been or is being carried out. Therefore, there is reason to be confident that better enzymes will be available in the near future and, consequently, there will be more applications of enzymatic modification of polar lipids.

With the expectation of better and cheaper enzymes, the task is therefore left to product developers and process engineers. Despite the work done so far, a lot more is needed for the development of new polar lipid products or their derivatives, to cater for new discoveries of the functional roles of PLs and other related polar lipids. Studies so far remain largely at the laboratory synthesis level. However, new ideas are continually producing new products, new reaction routes, and new reaction systems. Furthermore, there are technical developments in reactors, processes, large-scale operations, and other engineering issues. We believe enzymatic approaches for polar lipid modification will gain more industrial applications in the future.

Acknowledgement

Financial support from the Danish Research Council for Technology and Production (FTP), the Centre for Advanced Food Studies (LMC), the Technical University of Denmark and Danisco is acknowledged.

References

Adlercreutz, P, Virto, C, Persson, M, Vaz, S, Adlercreutz, D, Svensson, I and Wehtje, E (2001) Enzymatic conversion of polar lipids. Principles, problems and solutions. *Journal of Molecular Catalysis B: Enzymatic*, 11, 173–178.

Adlercreutz, D, Budde, H and Wehtje, E (2002) Synthesis of phosphatidylcholine with defined fatty acid in the *sn*-1 position by lipase-catalysed esterification and transesterification reaction. *Biotechnology and Bioengineering*, 78, 403–411.

Adlercreutz, P, Lyberg, A-M and Adlercreutz, D (2003) Enzymatic fatty acid exchange in glycerophospholipids. *European Journal of Lipid Science and Technology*, 105, 638–645.

Aurich, I, Durrschmidt, P, Hirche, F and Ulbrich-Hofmann, R (1997) Transesterification of alkylphosphate esters by phospholipase D. *Biotechnology Letters*, 19, 875–879.

Aurich, I, Dürrschmidt, P, Hirche, F and Ulbrich-Hofmann, R (2002) Production of octadecylphospho-L-serine by phospholipase D. *Biotechnology Letters*, 24, 585–590.

Benning, C and Ohta, H (2005) Three enzyme systems for galactoglycerolipid biosynthesis are co-ordinately regulated in plants. *Journal of Biochemistry*, 280, 2397–2400.

Bjorkling, F and Godtfredsen, SE (1988) New enzyme catalysed synthesis of monoacyl galactoglycerides. *Tetrahedron*, 44, 2957–2962.

Bueschelberger, H-G (2004) Lecithins. In: *Emulsifiers in Food Technology* (RJ Whitehurst, ed), Blackwell Publishing Ltd, Oxford, pp. 1–39.

Bugaut, M, Kuksis, A and Myher, JJ (1985) Loss of sterospecificity of phospholipase C and D upon introduction of a 2-alkyl group into rac-1,2-diacylglycero-3-phosphocholine. *Biochimica et Biophysica Acta*, 835, 304–314.

Buikstra, FPM, van der Kruis, AJMM and van der Heiden, PHN (1997) Heat stable oil-in-water emulsion stabilized by hydrolysates. USP 5650190.

Carlsson, A and Herslof, B (1998) Lipophilic carrier preparation. USP 5716639.

Carlsson, A, Delogu, M and Herslof, B (1997) Oil-in water emulsions. USP 5688528.

Carlsson, A, Herslof, B and Petrovic-Kallholm, S (2000) Bilayers preparations. USP 6022561.

Chmiel, O, Melachouris, N and Tritler, H (1999) Process for the interesterification of phospholipids. USP 5989599.

Chung, GH and Yang, YL (2004) Method for preparing lysophosphatidylethaolamine. USP 6773902.

Clausen, K (2001) Enzymatic oil-degumming by novel microbial phospholipase. *European Journal of Lipid Science and Technology*, 103, 333–340.

Coderch, L, Lopez, O, de la Maza, A and Parra, JL (2003) Ceramides and skin function. *American Journal of Clinical Research*, 4, 107–129.

Colarow, L, Masson, G and Truek, HU (1994) Heat stable oil and water emulsion and preparation thereof. USP 5314706.

Colombo, D, Ronchetti, F, Scala, A, Taino, IM, Albini, FM and Toma, L (1995) Optically pure 1-*O*-beta-D-glucosyl-sn-glycerols and 3-*O*-beta-D-glucosyl-sn-glycerols and galactosayl-sn-glycerols through lipase-catalyzed transformations. *Tetrahedron Letters*, 36, 4865–4868.

Colombo, D, Scala, A, Taino, IM, Toma, L and Ronchetti, F (1996) 1-*O*-, 2-*O*- and 3-*O*-β-D-glycosyl-sn-glycerols: Structure–anti-tumor promoting activity relationship. *Bioorganic & Medicinal Chemistry Letters*, 6, 1187–1190.

Colombo, D, Ronchetti, F, Scala, A and Toma, L (1998a) Bioactive glycoglycerolipid analogues: an expeditious enzymatic approach to mono- and diesters of 2-*O*-β-D-galactosylglycerol. *Tetrahedron: Asymmetry*, 9, 2113–2119.

Colombo, D, Scala, A, Taino, IM, Toma, L, Ronchetti, F, Tokuda, H, Nishino, H, Nagastu, A and Sakakibara, J (1998b) Inhibitory effects of fatty acid monoesters of 2-*O*-β-D-glucosylglycerol on Epstein-Barr virus activation. *Cancer Letters*, 123, 83–86.

Colombo, D, Compostella, F, Ronchetti, F, Scala, A, Tokuda, H and Nishino, H (2001) Diesters of glycosylglycerols active in cancer chemoprevention. *European Journal of Medical Chemistry*, 36, 691–695.

Colombo, D, Compostella, F, Ronchetti, F, Reza-Elahi, S, Scala, A, Toma, L, Aoi, W, Kuchide, M, Takayasu, J, Tokuda, H and Nishino, H (2002) Inhibitory effect of stabilized analogues of glycoglycerolipids on Epstein-Barr virus activation and mouse skin tumor promotion. *Cancer Letters*, 186, 37–41.

Colombo, D, Franchini, L, Toma, L, Ronchetti, F, Tanaka, R, Takayasu, J, Nishino, H and Tokuda, H (2006) Cyclic and branched acyl chain galactoglycerolipids and their effect on anti-tumor-promoting activity. *European Journal of Medical Chemistry*, 41, 1456–1463.

D'Arrigo, P and Servi, S (1997) Using phospholipases for phospholipid modification. *Trends in Biotechnology*, 15, 90–6.

D'Arrigo, P, de Ferra, L, Piergianni, V, Ricci, A and Scarcelli, D (1994) Phospholipase D from *Streptomyces* catalyses the transfer of secondary alcohols. *Journal of Chemical Society: Chemical Communications*, 14, 1709–1710.

de Roode, BM, Peters, SWPG, Fransen, MCR, van der Padt, A and Boom, RM (2001) Optimisation and downstream processing of the almond-β-glucosidase mediated glucosylation of glycerol. *Biotechnology and Bioengineering*, 72, 568–572.

de Roode, BM, Fransen, MCR, van der Padt, A and Boom, RM (2003) Perspectives for the industrial enzymatic production of glycosides. *Biotechnology Progress*, 19, 1391–1402.

Dittrich, N and Ulbrich-Hofmann, R (2001) Transphosphatidylation by immobilized phospholipase D in aqueous media. *Biotechnology and Applied Biochemistry*, 34, 189–94.

Dittrich, N, Haftendorn, R, Ulbrich-Hofmann, R (1998) Hexadecylphosphocholine and 2-modified 1,3-diacylglycerols as effectors of phospholipiase D. *Biochimica et Biophysica Acta*, 1391, 265–272.

Doig, SD and Diks, RMM (2003a) Toolbox for exchanging constituent fatty acids in lecithin. *European Journal of Lipid Science and Technology*, 105, 359–367.

Doig, SD and Diks, RMM (2003b) Toolbox for modification of the lecithin headgroup. *European Journal of Lipid Science and Technology*, 105, 368–76.

Dubber, M, Patel, A, Sadalapure, K, Aumuller, I and Lindhorst, TK (2006) Synthesis of functionalised amphiphilic glycoconjugates and glycoclusters. *European Journal of Organic Chemistry*, 5357–5366.

Egger, D, Wehtje, E and Adlercreutz, P (1997) Characterization and optimization of Phospholipase A2 catalyzed synthesis of phosphatidylcholine. *Biochimica et Biophysica Acta*, 1343, 76–84.

Eibl, H and Unger, C (1998) Phospholipids-selective drugs in cancer-therapy. *Proceedings of Social and Experimental Biology and Medicine*, 29, 358–358.

Elimelech, R. (2002) Process for large-scale preparation of sphingosines and ceramides. USP 6469148 B1.

Fogarty, WM and Benson, CP (1982) Measurment of glucoside transfer to glycerol by *Aspergilus niger* transglucosidase. *Biotechnology Letters*, 4, 61–64.

Frohman, MA and Morris, AJ (1999) Phospholipase D structure and regulation. *Chemistry and Physics of Lipids*, 98, 127–140.

Fujima, T, Frusawa, H, Minamikawa, H, Ito, K and Shimizu, T (2006) Elastic precursor of the transformation from glycolipids nanotube to vesicle. *Journal of Physics: Condensed Matter*, 18, 3089–3096.

Garcia-Junceda, E, Garcia-Garcia, JF, Bastida, A and Fernandez-Meyoralas, A (2004) Enzymes in the synthesis of bioactive compounds: the prodigious decades. *Bioorganic and Medicinal Chemistry*, 12, 1817–1834.

Geho, WB (1980) Lipid membrane structure for drug delivery and composition containing a lipid membrane drug system. EP 0009842.

Gill, I and Valivety, R (2000) Enzymatic glycosylation in plasticized glass phases: A novel and efficient route to O-glycosides. *Angewandte Chemie International Edition*, 39, 3804–3808.

Guo, Z, Vikbjerg, A F and Xu, X (2005). Enzymatic modification of phospholipids for functional applications and human nutrition. *Biotechnology Advances*, 23, 203–259.

Haraldsson, GG and Thorarensen, A (1999) Preparation of phospholipids highly enriched with n–3 polyunsaturated fatty acids by lipase. *Journal of the American Oil Chemists' Society*, 76, 1143–1149.

Härröd, M and Elfman, I (1995) Enzymatic synthesis of phosphatidylcholine with fatty acids, isooctane, carbon dioxide, and propane as solvent. *Journal of the American Oil Chemists' Society*, 72, 641–646.

Hassler, DF and Bell, RM (1993) Ceramidases: enzymology and metabolic roles. *Advances in Lipid Research*, 26, 49–57.

Heringdorf, DMZ, Himmel, HM and Jakobs, KH (2002) Sphingosylphosphorylcholine – biological functions and mechanisms of action. *Biochimica et Biophysica Acta – Molecular and Cell Biology of Lipids*, 1582, 178–189.

Hidehiko, T, Susumu, H and Yuko, I (1995) Mass preparation of ceramide from fishes and shells at low costs – comprises extracting sphingomyelin or ceramide 2-aminoethyl phosphonate and eliminating phosphonic acid. JP 07-265089-A.

Hirche, F and Ulbrich-Hofmann, R (1999) The interfacial pressure is an important parameter for the rate of phospholipase D catalysed reactions in emulsion systems. *Biochimica et Biophysica Acta*, 1436, 383–389.

Hirche, F and Ulbrich-Hofmann, R (2000) The interdependence of solvent, acceptor alcohol and enzyme source in transphosphatidylation by phospholipase D. *Biocatalysis and Biotransformation*, 18, 343–53.

Hirche, F, Koch, MHJ, König, S, Wadewitz, T and Ulbrich-Hofmann, R (1997) The influence of organic solvents on phospholipid transformations by phospholipase D in emulsion systems. *Enzyme and Microbial Technology*, 20, 453–461.

Hosokawa, M, Takahashi, K, Kikuchi, Y and Hatano, M (1995) Preparation of therapeutic phospholipids through porcine pancreatic phospholipase A2-mediated esterification and lipozyme-mediated acidolysis. *Journal of the American Oil Chemists' Society*, 72, 1287–1291.

Hosokawa, M, Shimatani, T, Kanada, T, Inoue, Y and Takahashi K (2000) Conversion to docosahexaenoic acid-containing phosphatidylserine from squid skin lecithin by phospholipase D-mediated transphosphatidylation. *Journal of Agricultural Food Chemistry*, 48, 4550–4.

Hronowski, LJJ, Szarek, WA, Hay, GW and Krebs, A (1991) Synthesis and characterization of 2-*O*-β-lactosylglycerol, 1,2-di-*O*-β-lactosyl-(*R*,*S*)-glycerols and 1,2,3-tri-*O*-β-lactosylglycerol. *Carbohydrate Research*, 219, 33–49.

Hu, Y and Walker, S (2002) Remarkable structural similarities between diverse glycosyltransferases. *Chemistry and Biology*, 9, 1287–1296.

Huang, C-H (2001) Mixed-chain phospholipids: structures and chain-melting behaviour. *Lipids*, 36, 1077–1097.

Ito, M, Kurita, T and Kita, K (1995) A novel enzyme that cleaves the *N*-acyl linkage of ceramides in various glycosphingolipids as well as sphingomyelin to produce their lyso forms. *Journal of Biological Chemistry*, 270, 24370–24374.

Ito, M, Kurita, T, Mitsutake, S and Kita, K (1999) Process for the preparation of sphingolipids and sphingolipid derivatives. EP 0940409A1.

Iwasaki, Y, Mizumoto, Y, Okada, T, Yamamoto, T, Tsutsumi, K and Yamane, T (2003) An aqueous suspension system for phospholipase D-mediated synthesis of PS without toxic organic solvent. *Journal of the American Oil Chemists' Society*, 80, 653–657.

Jenski, LJ, Zerouga, M and Stillwell, W (1995) Omega-3 fatty acid-containing liposomes in cancer therapy. *Proceedings of Social and Experimental Biology and Medicine*, 210, 227–233.

Jorasch, P, Warnecke, DC, Lindner, B, Zahringer, U and Heinz, E (2000) Novel processive and nonprocessive glycosyltransferases from *Staphylococcus aureus* and *Arabidopsis thaliana* synthesize glycoglycerolipids, glycophospholipids, glycosphingolipids and glycosylsterols. *European Journal of Biochemistry*, 267, 3770–3783.

Juneja, LR, Yamane, T and Shimizu, S (1989) Studies on enzymatic conversion of phospholipids. 5. Enzymatic method of increasing phosphatidylcholine content of lecithin. *Journal of the American Oil Chemists' Society*, 66, 714–7.

Juneja, LR, Taniguchi, E, Shimizu, S and Yamane, T (1992) Increasing productivity by removing choline in conversion of phosphatidylcholine to phosphatidylserine by phospholipase-D. *Journal of Fermentation and Bioengineering*, 73, 357–61.

Kates, M (1990) *Glycolipids, Phosphoglycolipids and Sulfoglycolipids*, Plenum Press, New York.

Kim, DS, Kim, SY, Lee, JE and Park, GC (2003) Skin whitening cosmetic composition containing sphingosine-1-phosphate and derivative there. KR 2003087448-A.

Kita, K, Kurita, T and Ito, M (1996) Fatty acid transfer (reversed hydrolysis) and fatty acid exchange of various glycosphingolipids by sphingolipid ceramide *N*-deacylase. In: *Proceedings of the XVIII International Carbohydrate Symposium*, Milan, Italy, p. 549.

Kita, K, Kurita, T and Ito, M (2001) Characterization of the reversible nature of the reaction catalyzed by sphingolipid ceramide *N*-deacylase. A novel form of reverse hydrolysis reaction. *European Journal of Biochemistry*, 268, 592–602.

Koeller, KM and Wong, C (2000) Synthesis of complex carbohydrates and glycoconjugates: enzyme-based and programmable one-pot strategies. *Chemical Reviews*, 100, 4465–4493.

Koketsu, M, Nitoda, T, Sugino, H, Juneja, LR, Kim, M, Yamamoto, T, Abe, N, Kajimoto, T and Wong, C-H (1997) Synthesis of a novel sialic acid derivative (sialylphospholipid) as an antirotaviral agent. *Journal of Medical Chemistry*, 40, 3332–3335.

Kokusho, Y, Tsunoda, A, Kato, S, Machida, H and Iwasaki, S (1993) Production of various phosphatidylsaccharides by phospholipase-D from *Actinomadura* sp. strain No 362. *Bioscience, Biotechnology and Biochemistry*, 57, 1302–5.

Laane, C, Boeren, S, Vos, K and Veeger, C (1986) Rules for optimization of biocatalysis in organic solvents. *Biotechnology and Bioengineering*, 30, 81–87.

Lafont, D, Carriere, F, Ferrato, F and Boullanger, P (2006) Syntheses of an α-D-Gal-(1→6)-β-D-Gal diglyceride, as lipase substrate. *Carbohydrate Research*, 341, 695–704.

Larsen, E, Kharazmi, A, Christensen, SB, Christensen, LP and Brandt, K (2006) Use of glycosides of mono and diacylglycerol as anti-inflammatory agents. USP 7084122 B2.

Lasson-Backstrom, C (1995) Phospholipids containing omega-3 fatty acids. USP 5434183.

Li, XC, Yu, LB, Zhang, H, Chu, SD and Guan HS (2003) The synthesis of glycoglycerolipids. *Chinese Chemical Letters*, 14, 776–778.

Lindberg, J, Svensson, S, Påhlsson, P and Konradsson, P (2002) Synthesis of galactoglycerolipids found in the HT29 human colon carcinoma cell line. *Tetrahedron*, 58, 5109–5117.

Mandal, SB, Sen, PC and Chakrabarti, P (1980) In vitro synthesis of phosphatidylinositol and phosphatidylcholine by phospholipase-D. *Phytochemistry* 19, 1661–1663.

Marino-Albernas, JR, Bittman, R, Peters, A and Meyhew, E (1996) Synthesis and growth inhibitory properties of 1-O-hexadecyl-2-O-methyl-sn-glycerol, analogs of the antitumor ether lipid ET-18-OCH_3 (Edelfosine). *Journal of Medicinal Chemistry*, 39, 3241–3247.

Matsufuji, M, Nagamatsu, Y and Yoshimoto, A (2000) Protective effects of bacterial glyceroglycolipid M874B cell death caused by exposure to heat and hydrogen peroxide. *Journal of Bioscience and Bioengineering*, 89, 345–349.

Michaelis, M, Geisslinger, G and Scholich, K (2004) Use of S1 P or functional fragments or derivatives thereof for the preparation of pharmaceutical compounds. EP 1481680A1.

Mingarro, I, Abad, C and Braco, L (1994) Characterization of acylating and deacylating activities of an extracellular phospholipase A2 in a water-restricted environment. *Biochemistry*, 33, 4652–4660.

Mitsutake, S, Kita, K, Okino, N and Ito, M (1997) Ceramide synthesis by sphingolipid ceramide *N*-deacylase: new assay for ceramidase activity detection. *Analytical Biochemistry*, 247, 52–57.

Morigaki, E, Miura, Y, Takahata, K, Tada, M, Nakajima, S and Baba, N (1999) Enzymatic preparation of sphingosine 1-phosphate. *Journal of Chemical Research-S*, 12, 774–775.

Morimoto, T, Nagatsu, A, Murakami, N and Sakakibara, J (1995) Chemoenzymatic synthesis of 1-*O*-acyl-3-*O*-(6'-*O*-acyl-*β*-D-galactopyranosyl)-sn-glycerol. *Tetrahedron*, 51, 6443–6450.

Nagao, A, Ishida, N and Terao, J (1991) Synthesis of 6-phosphatidyl-L-ascorbic acid by phospholipase-D. *Lipids*, 26, 390–4.

Nakagawa, T, Morotomi, A, Tani, M, Sueyoshi, N, Komori, H and Ito, M (2005) C18:3-GM1a induces apoptosis in Neuro2a cells: enzymatic remodelling of fatty acyl chains of glycosphingolipids. *Journal of Lipid Research*, 46, 1103–1112.

Nakajima, N, Furuya, T, Ishihara, K, Matsumura, S, Mizobata, S and Hamada, H (1997) Enzymatic production of *β*-thujaplicin 2-*O*-*β*-D-glucoside in a membrane reactor using an insoluble UDP-glucosyltransferase fraction from cultured cells of *Eucalyptus perriniana*. *Journal of Fermentation and Bioengineering*, 84, 455–460.

Nakano, H, Kiso, T, Okamoto, K, Tomita, T, Manan, MBA and Kitahata, S (2003) Synthesis of glycosylglycerol by cyclodextrin glucanotransferases. *Journal of Bioscience and Bioengineering*, 95, 583–588.

Okahata, Y and Mori, T (1998) Biocatalytic reactions in organic media by using lipid-coated enzymes. *Journal of Synthetic Organic Chemistry, Japan*, 11, 931–939.

Okahata, Y, Niikura, K and Ijiro, K (1995) Simple transphosphatidylation of phospholipids catalyzed by a lipid-coated phospholipase-D in organic solvents. *Journal of the Chemical Society-Perkin Transactions 1*, 7, 919–25.

Paltauf, F and Hermetter, A (1994) Strategies for the synthesis of glycerophospholipids. *Progress in Lipid Research*, 33, 239–328.

Park, CW, Kwon, SJ, Han, JJ and Rhee, JS (2000) Transesterification of phosphatidylcholine with eicosapentaenoic acid ethyl ester using phospholipase A2 in organic solvent. *Biotechnology Letters*, 22, 147–150.

Pedersen, KB (2001) Interesterification of phospholipids. USP 6284501.

Peng, L, Xu, X, Mu, H, Høy, C-E and Adler-Nissen, J (2002) Production of structured phospholipids by lipase-catalyzed acidolysis: optimization using response surface methodology. *Enzyme and Microbial Technology*, 31, 523–532.

Persson, M, Svensson, I, and Adlercreutz, P (2000) Enzymatic fatty acid exchange in digalactosyldiacylglycerol. *Chemistry and Physics of Lipids*, 104, 13–21.

Pisch, S, Bornscheuer, UT, Meyer, HH and Schmid, RD (1997) Properties of unusual phospholipids. 4. Chemoenzymatic synthesis of phospholipids bearing acetylenic fatty acids. *Tetrahedron*, 53, 14627–14634.

Pisvejcova, A, Rossi, C, Husakova, L, Kren, V, Riva, S and Monti, D (2006) *β*-1,4-Galactosltransferase catalysed glycosylation of sugar derivatives: Modulation of enzyme activity by *α*-lactalbumin, immobilization and solvent tolerance. *Journal of Molecular Catalysis B: Enzymatic*, 39, 98–104.

Ponting, CP and Kerr, ID (1996) A novel family of phospholipase D homologues that includes phospholipid synthases and putative endonucleases identification of duplicated repeats and potential active-site residues. *Protein Science*, 5, 914–22.

Qian, X, Sujino, K, Palcic, MM and Ratcliffe, RM (2002) Glycosyltransferases in oligosaccharide synthesis. *Journal of Carbohydrate Chemistry*, 21, 911–942.

Ramm, W, Scatton, W, Wagner-Dobler, I, Wray, V, Nimtz, M, Tokuda, H, Enjyo, F, Nishino, H, Beil, W, Heckmann, R, Lurtz, V and Lang, S (2004) Diglucosyl-glycerolipid from the marine sponge-associated *Bacillus pumilus* strain AAS3: their production, enzymatic modification and properties. *Applied Microbiology and Biotechnology*, 64, 497–504.

Rich, JO and Khmelnitsky, YL (2001) Phospholipase D-catalyzed transphosphatidylation in anhydrous organic solvents. *Biotechnology and Bioengineering*, 7, 2374–2377.

Romeo, A, Kirschner, G and Menon, G (1994) Use of phosphoryl-L-serine-*N*-acyl-sphingosine in compositions for treating nervous system disorders and diseases involving modified immunological reactivity, e.g., ischemia, trauma, Alzheimer's disease, etc. WO/1994/003178.

Rowan, AS and Hamilton, CJ (2006) Recent developments in preparative enzymatic syntheses of carbohydrates. *Natural Product Reports*, 23, 412–443.

Sakai, K, Okuyama, H, Yura, J, Takeyama, H, Shinagawa, N, Tsuruga, N, Kato, K, Miura, K, Kawase, K, Tsujimura, T, Naruse, T and Koike, L (1992) Composition and turnover of phospholipids and neutral lipids in human breast cancer and reference tissues. *Carcinogenesis*, 13, 578–584.

Schäffner, I, Rücknagel, KP, Mansfeld, J and Ulbrich-Hofmann, R (2002) Genomic structure, cloning and expression of two-phospholipase D isoenzymes from white cabbage. *European Journal of Lipid Science and Technology*, 104, 9–87.

Schneider, M (1997) Phospholipids. In: *Lipid Technologies and Applications* (FD Gunstone and FB Padley, eds), Marcel Dekker Inc., New York, pp. 51–78.

Servi, S (1999) Phospholipases as synthetic catalysts. *Topic of Current Chemistry*, 200, 127–58.

Shigematsu, Y and Hasegawa, M (1989) Process for producing egg yolk lecithin reduced PE content and/or containing substantially no impurities. USP 4847015.

Silhavy, TJ and Boos, W (1973) A convenient synthesis of (2*R*)-glyceryl-β-D-galactopyranoside. *Journal of Biological Chemistry*, 248, 6571–6574.

Smeets, JWH, De Pater, RM and Lambers, JWJ (1997) Enzymatic synthesis of ceramides and hybrid ceramides. USP 5610040.

Song, JK, Han, JJ and Rhee, JS (2005) Phospholipases: Occurrence and production in microorganisms, assay for high-throughput screening, and gene discovery from natural and man-made diversity. *Journal of the American Oil Chemists' Society*, 82, 691–705.

Susumu, H, Hidehiko, T and Yumiko, T (1999) Process for preparing sphingomyelin and ceramide from erythrocyte as a starting material and a curing agent or cosmetic formulated with ceramide. USP 5912152.

Svensson, I, Adlercreutz, P and Mattiasson, B (1992) Lipase-catalyzed acidolysis of phosphatidylcholine at controlled water activity. *Journal of the American Oil Chemists' Society*, 69, 986–991.

Takahashi, K and Hosokawa, M (2001) Production of tailor-made polyunsaturated phospholipids through bioconversion. *Journal of Liposome Research*, 11, 343–353.

Takami, M and Suzuki, Y (1994a) Synthesis of novel phosphatidyldihydroxyacetone via transphosphatidylation reaction by phospholipase D. *Bioscience, Biotechnology and Biochemistry*, 58, 2136–9.

Takami, M and Suzuki, Y (1994b) Enzymatic synthesis of novel phosphatidylgenipin and its enhanced cytotoxicity. *Bioscience, Biotechnology and Biochemistry*, 58, 1897–1898.

Takami, M and Suzuki, Y (1995) Transphosphatidylation reaction of phosphatidylcholine to 4-methoxyphenol in water-immiscible organic-solvents with immobilized phospholipase-D. *Journal of Fermentation and Bioengineering*, 79, 313–6.

Takami, M, Hidaka, N, Miki, S and Suzuki, Y (1994) Enzymatic-synthesis of novel phosphatidylkojic acid and phosphatidylarbutin, and their inhibitory effects on tyrosinase activity. *Bioscience, Biotechnology and Biochemistry*, 58, 1716–7.

Takenaka, F and Uchimaya, H (2000) Synthesis of α-D-glucosylglycerol by α-glucosidase and some of its characteristics. *Bioscience, Biotechnology and Biochemistry*, 64, 1821–1826.

Thiem, J and Gambert, U (1997) Chemical transformations employing glycosyltransferases. *Topics in Current Chemistry*, 186, 21–43.

Tremblay, PA, Marziani, F, Tino, JAF and Pilkiewicz, FG (1990) Enzymatic synthesis of soluble phosphatides from phospholipids. WO/1991/016444.

Ulbrich-Hofmann, R, Lerchner, A, Oblozinsky, M and Bezakova, L (2005) Phospholipase D and its application in biocatalysis. *Biotechnology Letters*, 27, 535–543.

Vaz, S, Persson, M, Svensson, I and Adlercreutz, P (2000) Lipase catalysed fatty acid exchange in digalactosyldiacylglycerol. Improvement of the yield due to the addition of phenylboronic acid. *Biocatalysis and Biotransformation*, 18, 1–12.

Vessey, DA, Kelley, M and Karliner, JS (2005) A rapid radioassay for sphingosine kinase. *Analytical Biochemistry*, 337, 136–142.

Vikbjerg, AF, Mu, H and Xu, X (2005) Parameters affecting incorporation and by-product formation during the production of structured phospholipids by lipase-catalyzed acidolysis in solvent-free system. *Journal of Molecular Catalysis B-Enzymatic*, 36, 14–21.

Vikbjerg, AF, Rusig, J-Y, Jonsson, G, Mu, H and Xu, X (2006a) Strategies for production and purification of lipase-catalyzed structured phospholipids. *European Journal of Lipid Science and Technology*, 108, 802–811.

Vikbjerg, AF, Mu, H and Xu, X (2006b) Elucidation of acyl migration during lipase-catalyzed production of structured phospholipids, *Journal of the American Oil Chemists' Society*, 83, 609–614.

Vikbjerg, AF, Mu, H and Xu, X (2007) Synthesis of structured phospholipids by immobilized PLA2 catalyzed acidolysis. *Journal of Biotechnology*, 128, 545–554.

Virto, C and Adlercreutz, P (2000) Two-enzyme system for the synthesis of 1-lauroyl-rac-glycerophosphate (lysophosphatidic acid) and 1-lauroyl-dihydroxyacetonephosphate. *Chemistry and Physics of Lipids*, 104, 175–184.

Waite, M (1999) The PLD superfamily: insights into catalysis. *Biochimica et Biophysica Acta*, 1439, 187–197.

Wandzik, I, Bugla, J and Szeja, W (2006) Regio- and stereoselectivity in preparation of 2-deoxysugar glycoglycerolipid derivatives evaluated by high resolution ^{1}H NMR. *Chemistry and Physics of Lipids*, 139, 77–83.

Wang, X (2000) Multiple forms of phospholipase D in plants: the gene family, catalytic and regulatory properties, and cellular functions. *Progress of Lipid Research*, 39, 109–149.

Wang, P, Schuster, M, Wang, YF and Wong, CH (1993) Synthesis of phospholipid inhibitor conjugates by enzymatic transphosphatidylation with phospholipase-D. *Journal of American Chemical Society*, 115, 10487–10491.

Wang, X, Xu, L and Zheng, L (1994) Cloning and expression of phosphatidylcholine-hydrolyzing phospholipase D from Rincinus communis L. *Journal of Biological Chemistry*, 269, 20312–20317.

Weil, MJ, Zang, Y and Nair, MG (2005) Tumor cell proliferation and cyclooxygenase inhibitory constituents in horseradish (*Armoracia rusticana*) and wasabi (*Wasabia japonica*). *Journal of Agricultural Food Chemistry*, 53, 1440–1444.

Yamamoto, Y, Hosokawa, M and Miyashita, K (2006) Production of phosphatidylcholine containing conjugated linoleic acid mediated by phospholipase A_2. *Journal of Molecular Catalysis B-Enzymatic*, 41, 92–96.

Yamane, T, Juneja, LR, Li, D and Shimizu, S (1990) Enzymatic preparation of phospholipids containing various polar head groups. *Annals of New York Academy for Science*, 613, 686–690.

Yaqoob, M, Nabi, A and Masoom-Yasinzai, M (2001) Bioconversion of phosphatidylcholine to phosphatidylserine using immobilized enzyme mini-columns. *Process Biochemistry*, 36, 1181–1185.

Yasutake, N, Totani, K, Harada, Y, Haraguchi, S, Murata, T and Usui, T (2003) Efficient synthesis of glyceroyl β-lactoside and its derivatives trough a condensation reaction by cellulase. *Biochimica et Biophysica Acta*, 1620, 252–258

Yesair, DW (1998) Method for making lysophosphatidylcholine. USP 5716814.

Zambonelli, C, Morandi, P, Vanoni, MA, Tedeschi, G, Servi, S, Curti, B, Carrea, G, Di Lorenzo, R and Monti, D (2003) Cloning and expression in *Escherichia coli* of the gene encoding *Streptomyces* PMF PLD, a phospholipase D with high transphosphatidylation activity. *Enzyme and Microbial Technology*, 33, 676–688.

Zerouga, M, Stillwell, W and Jenski, LJ (2002) Synthesis of a novel phosphatidylcholine conjugated to docosahexaenoic acid and methotrexate that inhibits cell proliferation. *Anticancer Drugs*, 13, 301–311.

Zhang, L, Hellgren, LI and Xu, X (2006) Enzymatic production of ceramide from sphingomyelin. *Journal of Biotechnology*, 123, 93–105.

Zhang, L, Hellgren, LI and Xu, X (2007) Immobilization of phospholipase C for the production of ceramide from sphingomyelin hydrolysis. *Journal of the American Oil Chemists' Society*, 84, 237–247.

CHAPTER 4

Uses of phospholipids as functional ingredients

ERNESTO HERNANDEZ[1] AND NATHALIE QUEZADA[2]

[1]Omega Pure, Houston, Texas, USA and [2]Texas A&M University, College Station, Texas, USA

A. Introduction

Commercial phospholipids, or lecithin, are mainly by-products from the processing of soybean oil. Lecithin is widely used in the food, cosmetic and pharmaceutical industries because of its numerous surfactant and bioactive properties. Most lecithin is obtained from the process of degumming of crude soybean and other vegetable oils. The main reasons for degumming soybean oil are: 1) to prepare the oil for caustic or physical refining; 2) to avoid settling of phospholipid as a gum during storage and transporting; 3) to produce lecithin. Degumming is usually the first step in processing soybean oil, and it consists of the addition of water to the crude oil to hydrate the more hydrophilic lecithin, which then separates in a distinct phase, thus allowing its removal by centrifugation. Crude soybean oil is the main source of all commercial lecithin.

Lecithin is a mixture of several phospholipids and is commonly present in cells and plant membranes (see Chapter 2). Another minor commercial source of lecithin is derived from egg yolk. Lecithin is usually defined as a mixture of acetone-insoluble polar lipids and triacylglycerols, together with other minor components, produced during degumming of crude vegetable oils and animal fats (Szuhaj, 1988). The production of lecithin in the US in 2004 was estimated at close to 96,000 tons.

Phospholipids generally contain a glycerol or a sphingosyl backbone and are divided into four main classes: glycerophospholipids, sphingolipids, ether phospholipids and phosphonopholipids, depending on their backbones and bonding types (Akoh & Min, 1998). Glycerophospholipids are built up from glycerol, fatty acids, phosphoric acid and a second hydroxyl compound which

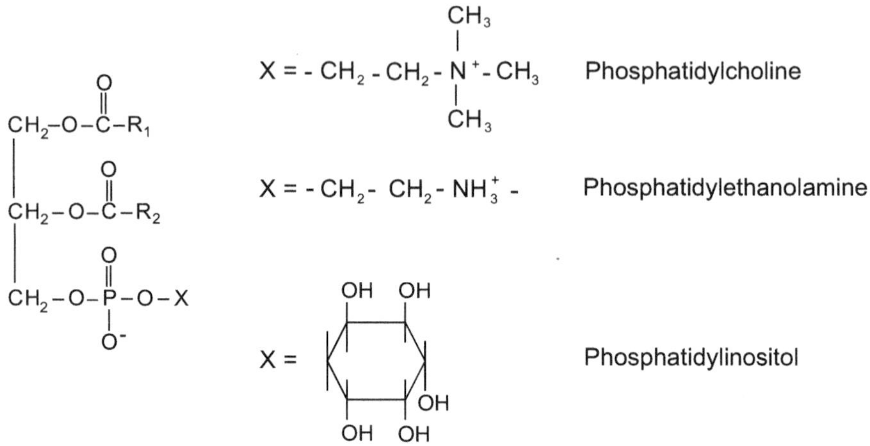

Figure 4.1. Structure of glycerophospholipids (see Chapter 1). R_1, R_2 = alkyl groups of fatty acids.

Table 4.1. Composition (%) of commercial lecithin (see Chapter 2)

	Regular	De-oiled
Triacylglycerols	35	3
Phosphtaidylcholine	16	24
Phosphatidylethanolamine	14	20
Phosphatidylinositol	10	14
Phosphatidic acid	5	7
Phytoglycolipids	11	15
Carbohydrates	5	8
Moisture	1	1

commonly contains nitrogen. The hydroxyl group can be choline, ethanolamine, serine, inositol and glycerol, as illustrated in Figure 4.1.

The typical composition of commercial lecithin is shown in Table 4.1 (Brekke, 1978). The gums or wet lecithin resulting from straight centrifugation contain approximately 50% water, which has to be quickly removed in order to avoid microbial spoilage. Bleaching agents such hydrogen peroxide are usually added before drying to less than 1% moisture. Commercial lecithins have been classified in at least six different grades which are divided into natural, refined and chemically modified (Table 4.2).

Table 4.1 shows two types of lecithins sold commercially. De-oiled lecithin is produced by removing residual oil with a solvent, usually acetone. Alcohol fractionation is also done with oil-free lecithin to produce a fraction high in

Table 4.2. Classification of lecithins

1.	Natural
	Fluid
	Plastic
2.	Refined
	Blended
	Oil-free
	Fractionated
3.	Chemically modified

phosphatidylcholine (alcohol-insoluble) and a fraction high in phosphatidylinositol (alcohol-soluble). Commercial chemically modified lecithins include hydrogenated, hydroxylated, acetylated, sulfonated and halogenated products.

Lecithin is considered GRAS (Generally Recognized As Safe) by the US Food and Drug Administration (FDA) under 21 CFR 184.1400. Chemically modified lecithins require special labelling; i.e., if hydroxylated, lecithin should be identified as 'Hydroxylated Soy Lecithin' or 'Hydroxylated Lecithin'. When enzymatically modified, the phrase 'Enzymatically Modified Lecithin' should be included on labelling. These modifications modify emulsification properties and improve water dispersibility.

B. Properties of phospholipids

Phospholipids have a bipolar, amphiphilic molecular structure due to a lipophilic part in the form of two fatty acids and a hydrophilic group in the form of a phosphoric acid based esters. Because of this amphiphilic character, phospholipids show a tendency to form bilayers or micelles and liposomes. As a result, they find many applications in foods, cosmetic and pharmaceutical products as natural emulsifiers, wetting and dispersing agents (Szuhaj, 1983). Also, phospholipids tend to form bilayer structures or micelles which make them able to encapsulate specific reactive agents into liposomes (Figure 4.2) (Bretscher, 1985).

The dispersion and emulsion stabilization properties of lecithins have been directly associated with the charge of the phospholipids (Rydhag & Wilton, 1981), as well as the composition of the fatty acids (Kauffman *et al.*, 1990).

The emulsification properties of lecithins are often characterized by their hydrophilic–lipophilic balance (HLB) value. This value varies between 1 and 20 where 1 is assigned to a surfactant that is oil-soluble and a value of 20 to emulsifiers that are completely soluble in water. Typical lecithins with 62% acetone-insolubles have an HLB of approximately 4. Oil-free or granulated

Lecithins are used as stabilizers and emulsifiers in many other food products such as baked goods, cheeses, meat and poultry products, dairy and imitation dairy products. (Brekke, 1978; Szuhaj, 1980, 1983). Typical levels of lecithins used in foods vary between 0.2 and 1.0%. Table 4.3 shows some food applications of phospholipids.

D. Pharmaceutical applications of phospholipids

Phospholipids play a significant role in maintaining cell structure integrity and in functions such as membrane permeability, membrane fluidity, membrane interactions (lipid–protein) and membrane deterioration (White *et al.*, 2001). Phospholipids are also reported to be involved in metabolism-related diseases (Lohmeyer & Bittmann, 1994; Lee, 1998) and as signalling compounds in the regulation of basic biological processes (Izumi & Shimizu, 1995; Hannun *et al.*, 2001). They have also been reported to help in the prevention of some neurological diseases (Schneider, 2001; Guo *et al.*, 2005). Phospholipids have been used as nutritional supplements because they have shown beneficial physiological effects. Soy phospholipid mixtures have been shown to reduce elevated blood cholesterol and triacylglycerols (Knuiman *et al.*, 1989; Wilson *et al.*, 1998). Soy phosphatidylcholine helps in liver detoxification and repair of damaged liver tissue (Lieber *et al.*, 1990) and phosphatidylserine from soybean, brain and other sources improves cognitive functions (Pepeu *et al.*, 1996). Egg phospholipids have been used in infant formula to improve visual acuity (Koletzko *et al.*, 1995).

Phospholipids, along with proteins, are the major structural components of membranes of most kinds of cells, including bacteria and mammalian tissues. The key to the biological role of phospholipids is the ability of the phosphate group to combine with water and hydrophilic molecules such as proteins, sugars and oxygen, while the fatty acids combine with hydrophobic substances, such as hormones and non-esterified fatty acids (Bretscher, 1985).

When ingested, phospholipids act as emulsifiers in the bile digestive fluid, ensuring fine dispersion of fatty food molecules in the water phase, thus improving digestion and absorption. Phospholipids combined with bioactive substances can enhance their bioavailability such as in the case of long-chain polyunsaturated fatty acids (Carnielli *et al.*, 1998), *n*–3 polyunsaturated fatty acids (Cansell *et al.*, 2003, 2006) and tocopherol (Nacka *et al.*, 2001). Phospholipids act also as surface-active wetting agents in the lungs (Robertson *et al.*, 1990), intestines and kidneys; phospholipids ensure miscibility at air–liquid and liquid–solid interfaces.

Normal intake of lecithin in humans is reported at 6 g/day because of lower fat foods consumption (Zeisel *et al.*, 1980). Choline is an essential precursor in the synthesis of the neurotransmitter acetylcholine and its deficiency in the human diet has been associated with impaired renal function, infertility, bone abnormalities and hypertension. PS has been shown to enhance neuronal membrane function and has been prescribed to elderly people to improve cognitive function (Crook *et al.*, 1991). PS is found at levels of up to 1% in commercial lecithin from vegetable oils and 3% in commercial lecithin from eggs. The FDA authorized two health claims for PS on February 24, 2003. The first claim was "Phosphatidylserine (PS) may reduce the risk of cognitive dysfunction in the elderly", and the second claim was "Phosphatidylserine (PS) may reduce the risk of dementia in the elderly". Phospholipids, as structured phospholipids, are also used to enhance the biological activity of other functional lipids such as conjugated linoleic acid (Hossen & Hernandez, 2005).

Intravenous administration of oil-in-water microemulsions that are pharmaceutically acceptable has also been reported. An example of this is a microemulsion composed of medium-chain triacylglycerol (MCT), soybean phosphatidylcholine and poly(ethylene glycol)(660)-12-hydroxystearate (12-HSA- EO15) as amphiphiles, and poly(ethylene glycol) 400 (PEG 400) and ethanol as co-solvents. The mean droplet diameter of the oil-in-water emulsion reported was between 60 and 200 nm. It was concluded that it is possible to administer the microemulsion without producing any significant effect on acid–base balance, blood gases, plasma electrolytes, mean arterial blood pressure or heart rate (concentrations of up to 0.5 ml/kg of the microemulsion) (von Corswant *et al.*, 1998).

E. Cosmetic applications of phospholipids

Phospholipids, along with a wide range of oils and fats, are currently being used as inert and bioactive ingredients in many cosmetic applications. Typical applications of lipids in cosmetics include emollients and specific ingredients for skin care and treatment, hair care, make-up and decorative products. Other commonly used lipids in cosmetics include emulsifiers and structured lipids in which phospholipids can be used used as coadjuvants. Typical properties assigned to cosmetic products include skin moisturization, emolliency and spreadability and coadjuvants to help maintain skin pliability and softness (Bucalo, 1998). The role of lipid cosmetics, besides moisturization, has expanded to transdermal effects for the transport of various skin barrier modulators and other bioactive components and the use of emulsifiers and surfactants. The cosmetics industry is estimated at over $25 billion in the US.

One general rationale for the use of certain lipids in cosmetics is the belief that compounds with a similar composition to human skin will be compatible with and may have beneficial properties for the epidermis. The major components of sebaceous lipids are 45–60% triacylglycerols, 25% wax and sterol esters, 12–15% squalene and 10% free fatty acids (Lampe *et al.*, 1983). The outer layer of the skin, or stratum corneum, has an abundance of triacylglycerols and fatty acids, along with other components such as cholesterol and ceramides (Norlen, 2003). Also, as mentioned above, the cell walls contain phospholipids. This has been applied in the design of high emolliency formulations using oils rich in oleic, linoleic and palmitic acids, such as olive, corn and palm oils, which are widely used in cosmetics. These formulations use phospholipids as one of the emulsifying agents.

Lipids are the major components in cosmetics applications, such as skin and body care products. However, lipids have also been used as vehicles or carriers for penetration enhancement in dermatologic delivery systems specifically for applications such as skin repair. In some cases, the lipids may have a combined cosmetic or pharmaceutical effect. For example, in addition to emolliency, some products are marketed as having anti-wrinkle and skin protection properties. Medium-chain and short-chain monoacylglycerols have been proven to be effective against microorganisms found on skin that may trigger infections. Phytosterols are shown to have a positive impact on irritated human skin. The term 'cosmeceutical' has been coined to define cosmetic products with medical properties. Also, oils rich in essential fatty acids have been reported to suppress human granulocyte elastase, a tissue-degenerative enzyme that is released when tissue inflammation occurs (Ghosh *et al.*, 1995). Oils rich in essential fatty acids, especially from animal origin, have been reported to suppress inflammation. As in pharmaceutical applications, the cosmetic properties of lipids may be increased synergistically when used in combination with phospholipids.

The number of cosmetic products with bioactive ingredients is constantly increasing. For example, cosmetic blends of natural tocopherols with MCT are being commercialized. These products are reported to have quick skin penetration, and high oxidative stability and solubilization. One of the most common fatty acids used in cosmetics is oleic acid; it is highly stable to oxidation and is reported to have high emolliency. The use of structured lipids has also been increasing in the pharmaceutical and cosmetics industries. The combination of short-chain fatty acids (caprylic and capric acid) and long-chain polyunsaturated fatty acids (such as long-chain n–3 fatty acids) allows for the design of synthetic lipids for specific applications (Jenning *et al.*, 2000).

The outer layer of the skin can be a difficult barrier to penetrate. Therefore permeation-enhancing agents may be necessary to allow intake of some drugs.

There are several advantages reported for transdermal drug delivery: 1) bypassing of gastrointestinal and hepatic metabolism, 2) the potential for enhanced efficacy, 3) lower side effects through a more controlled delivery rate, and 4) increasing patient compliance through easier application and dosing frequency (Charman *et al.*, 1997).

One method for increasing permeation through the skin barrier is the use of vesicular systems, such as liposomes and niosomes (non-ionic surfactant vesicles). The effectiveness of this method will depend on the physicochemical properties, composition, size and type of the vesicular system used for topical/transdermal delivery (Choi & Maibach, 2005). Phospholipid vesicles or liposomes have been examined as a means for transdermal delivery of bioactive and pharmaceutical compounds. A more flexible type of vesicle called a transferosome, made with phosphatidylcholine and sodium cholate, has been reported to be more effective in overcoming the epidermis barrier, due to its deformation ability which allows it to go through the lipid matrix of the stratum corneum (Schatzlein & Cebc, 1995).

F. Modified phospholipids

Phospholipids can be modified to improve their properties or to meet particular functional requirements. For this purpose, the fatty acids or the head group can be changed or modified. These modifications can be performed by physical, chemical and enzymatic modification (Doig & Dick, 2003) (see Chapters 2, 3 and 6).

Structured phospholipids are phospholipids in which the head group and/or fatty acid composition have been changed. Structured phospholipids have been synthesized with diverse fatty acids such as oleic acid (Yagi *et al.*, 1990), heptadecanoic acid (Svensson *et al.*, 1992), caprylic acid and eicosapentaenoic acid (Peng *et al.*, 2002), docosahexaenoic acid (Peng *et al.*, 2002; Hosokawa *et al.*, 1998), conjugated linoleic acid (Hossen & Hernandez, 2005; Peng *et al.*, 2002) and with phytosterols (Hossen & Hernandez, 2004). Structured phospholipids with conjugated linolenic acid (CLA) may be desirable for the development of 'nutraceutical lipids' which may improve health, and prevent or treat disease (Hossen & Hernandez, 2005).

Physical processes such as solvent fractionation using acetone and alcohol and chromatography have been used to obtain de-oiled and high purity phospholipids from commercial sources (Doig & Dicks, 2003). The chemical modification of phospholipids usually allows little or no control over the positional distribution of fatty acids in the final product (Willis *et al.*, 1998), and requires high temperatures, solvents and anhydrous conditions. All this makes the process difficult to use in food or nutritional applications (Osborn & Akoh, 2002). Enzymatic modification offers a great advantage in catalysed interesterification reactions, because the enzymes can control the positional distribution of fatty acids in the final product, due to their selectivity and regiospecificity (Akoh & Min, 1998). Use of enzymatic modification of phospholipids (Chapter 3) can used to increase the nutritional and functional properties of phospholipids as well (Guo *et al.*, 2005).

References

Akoh, C and Min, D (eds) (1998) *Food Lipids: Chemistry, Nutrition and Biotechnology*, Marcel Dekker, New York.

Brekke, OL (1978) In: *Handbook of Soy Oil Processing and Utilization* (DR Erickson *et al.* ed) AOCS Press, Champaign, IL.

Bretscher, M (1985) The molecules of cell membrane. *Scientific American*, 253, 100–108.

Bucalo, AF (1998) State of the industry. *Global Cosmetic Industr.*, July, pp. 32–36.

Carnielli, V, Verlato G, Perderzini, F, Luijendijk, I, Boerlage, A, Pedrotti, D and Sauer, P (1998) Intestinal absorption of long-chain polyunsaturated fatty acids in preterm infants fed breast milk or formula. *Am. J. Clin. Nutr.*, 67, 97–103.

Cansell, M, Nacka, F and Combe, N (2003) Marine lipid-based liposomes increase in vivo fatty acid bioavailability. *Lipids*, 38, 551–559.

Cansell, M, Moussaoui, N, Petit, A, Denizot, A and Combe, N (2006) Feeding rats with liposomes or fish oil differently affects their lipid metabolism. *Eur. J. Lipid Sci. Technol.*, 108, 6, 459–467.

Charman, WN, Porter CJH, Mithani, S and Dressman, JB (1997) Physicochemical and physiological mechanisms for the effects of food on drug absorption: The role of lipids and pH. *J. Pharm. Sci.*, 86, 269–404.

Choi, MJ and Maibach, HI (2005) Liposomes and niosomes as topical drug delivery systems. *Skin Pharmacol. Physiol.*, 18, 209–219.

Crook, TH., Tinklenberg, J, Yesavage, W, Petrie, MG, Nunzi, MG and Massari, DC (1991) *Neurology*, 41, 644–649.

Doig, S and Diks, R (2003) Toolbox for exchanging constituent fatty acids in lecithins. *Eur. J. Lipid Sci. Technol.*, 105, 359–367.

Ghosh, P, Whitehouse, M, Dawson, M and Turner, AG (1995) Anti-inflammatory composition derived from emu oil. USP 5431924.

Guo, Z, Vikbjerg, A and Xu, X (2005) Enzymatic modification of phospholipids for functional applications and human nutrition. *Biotechnol. Adv.*, 23, 203–259.

Hannun, Y, Luberto, C and Argraves, K (2001) Enzymes of sphingolipid metabolism: from modular to integrative signaling. *Biochemistry*, 40, 4893–4903.

Hossen, M and Hernandez, E (2004) Phospholipase D-catalyzed synthesis of novel phospholipid-phytosterol conjugates. *Lipids*, 39, 777–782.

Hossen, M and Hernandez, E (2005) Enzyme-catalyzed synthesis of structured phospholipids with conjugated linoleic acid. *Eur. J. Lipid Sci. Technol.*, 107, 730–736.

Izumi, T and Shimizu, T (1995) Platelet-activating factor receptor: gene expression and signal transduction. *Biochim. Biophys. Acta.*, 1259, 317–333.

Jenning, V, Thünemann, VF and Gohla, SH (2000). Characterization of a novel solid lipid nanoparticle carrier system based on binary mixtures of liquid and solid lipids *Int. J. Pharm.*, 199, 20, 167–177.

Kaufmann, P, Olsson, U and Harslof, BG (1990) A chemometric strategy for the classification of lecithins according to process performance. *J. Am. Oil Chem. Soc.*, 67, 537.
Knuiman, J, Beynen, A and Katan, M (1989) Lecithin intake and serum cholesterol. *Am. J. Clin. Nutr.*, 49, 266.
Koletzko, B, Edenhofer, S, Lipowsky, G and Reinhardt, D (1995) Effects of a low birthweight infant formula containing human milk levels of docosahexaenoic acid and arachidonic acid. *J. Pediatr. Gastroenterol. Nutr.*, 21, 200–208.
Lampe, MA, Williams, ML and Elias, PM (1983) Human epidermal lipids: characterization and modulations during differentiation. *J. Lipid Research*, 24, 131–140.
Lee, T (1998) Biosynthesis and possible biological functions of plasmogens. *Biochim. Biophys. Acta*, 1394, 129–145.
Lieber, C, De Carli, K, Mak, K, Kim, C and Leo, M (1990) Attenuation of alcohol hepatic fibrosis by polyunsaturated lecithin. *Hepatology*, 12, 1390–1398.
Lohmeyer, M and Bittmann, R (1994) Antitumor effect lipids and alkylphosphocholines. *Drug Future*, 19, 1021–1037.
Nacka, F, Cansell, M, Meleard, P and Combe, N (2001) Incorporation of α-tocopherol in marine lipid-based liposomes: *in vitro* and *in vivo* studies. *Lipids*, 36, 1313–1320.
Norlén, L (2003) Skin barrier structure function and formation. *Int. J. Cosmetic Sci.*, 25, 209–226.
Osborn, H and Akoh, C (2002) Structured lipids – novel fats with medical, nutraceutical, and food applications. *Comp. Rev. Food Sci. Food Safety*, 3, 93–103.
Peng, L, Xu, X, Mu, H, Hoy, C and Nissen, A (2002) Production of structured phospholipids by lipase-catalyzed acidolysis: optimization using response surface methodology. *Enzyme Microb. Technol.*, 31, 523–532.
Pepeu, G, Pepeu, L and Amaducci, L (1996) A review of phosphatidylserine pharmacological and clinical effects. Is phosphatidylserine a drug for the ageing brain? *Pharmacol. Res.*, 33, 73–80.
Robertson, B, Curstedt, T, Johansson, J, Jornvall, H and Kobayashi, T (1990) Structural and functional characterization of porcine surfactant isolated by liquid-gel chromatography. *Prog. Respir. Res.*, 25, 237–46.
Rydhag, L and Wilton, L (1981) The function of phospholipids of soybean lecithin in emulsions. *J. Am. Oil Chem. Soc.*, 58, 830.
Schatzlein, A and Cebc, G (1995) Skin penetration by phospholipids vesicles transfersomes. In: *Phospholipids: Characterization, Metabolism, and Novel Biological Applications* (G Cevc and F Paltauf, eds), AOCS Press, Champaign IL, pp. 188–207.
Schneider, M (2001) Phospholipids for functional food. *Eur. J. Lipid Sci. Technol.*, 103, 98–101.
Servi, S (1999) Phospholipases as synthetic catalysts. *Top. Curr. Chem.*, 200, 127–157.
Sinram, R (1991) The added value of specialty lecithins. *Oil Mill Gazetteer*, Sept., 22–26.
Svensson, I, Adlercreutz, P and Mattiasson, B (1992) Lipase-catalyzed acidolysis of phosphatidylcholine at controlled water activity. *J. Am. Oil Chem.*, Soc., 69, 986–991.
Szuhaj, B (1983) Lecithin production and utilization. *J. Am. Oil Chem. Soc.*, 60, 306–309.
Szuhaj, B (1988) *Lecithins: Sources, Manufacture and Uses*, AOCS Press, Champaign, Illinois.
Szuhaj, B and Nieuwenhuyzen, W (1998) Effects of lecithins and proteins on the stability of emulsions. *Fett/Lipid*, 100, 282–291.
Touitou, E, Godin, B and Wiess, C (2000) *Drug Development Research*, 50, 406–415.
von Corswant, C, Thorén, P and Engström, S (1998) *J. Pharm. Sci.*, 87(2), 200–208.
White, S, Ladokhin, A, Jayasinghe, S and Hristova, K (2001) How membranes shape protein structure. *J. Biol. Chem.* 276, 32395–32398.

Willis, W, Lencki, R and Marangoni, A (1998) Lipid modification strategies in the production of nutritionally functional fats and oils. *Crit. Rev. Food Sci. Nutrit.*, 38, 639–674.

Wilson, T, Meservey, C and Nicolosi, R (1998) Soy lecithin reduces plasma lipoprotein cholesterol and early atherogenesis in hypercholesterolemic monkeys and hamsters: beyond linoleate. *Atherosclerosis*, 140, 147–153.

Yagi, T, Nakanishi, T, Yoshizawa, Y and Fukui, F (1990) The enzymatic acyl exchange of phospholipids with lipases. *J. Ferment. Bioeng.*, 69, 23–25.

Zeisel, SH, Growdon, JH, Wurtman, RJ, Magil, SG and Logue, M (1980) Normal plasma choline responses to ingested lecithin. *Neurology*, 30, 1226.

CHAPTER 5

Physical properties of phospholipids

ANDERS CARLSSON

MediGelium AB, Stockholm, Sweden

A. Introduction

The many remarkable physical properties of phospholipids in water can be related to the tendency of the acyl chains to avoid contact with water, at the same time as the polar head group tends to be strongly hydrated. The adsorption of phospholipids at interfaces between aqueous solutions and air, another liquid phase, or a solid phase is one consequence of this. The extensive aggregation into micelles and liquid crystalline phases is another (Lindman & Wennerström, 1980).

The physical chemistry of polar lipids, including phospholipids, has been thoroughly reviewed in a number of textbooks and in individual chapters (see, for example, Larsson, 1986, 1994; Small, 1986; Cevc, 1993). In this chapter, some of the most important physical properties of phospholipids, both of synthetic and natural origin, will be presented. The main focus is on double-chain phospholipids, (i.e., 1,2-diacylphospholipids), but single-chain phospholipids or lysophospholipids are included. Special attention is paid to the phase behaviour of phospholipids in water.

The most common 1,2-diacylphospholipid, phosphatidylcholine (PC), of which lecithin is a crude form, is a polar, bilayer-forming lipid used in many applications, particularly as an emulsifier in food and pharmaceutical emulsions. In addition, due to its bilayer-forming ability, it may be used for the formation of different kinds of aggregates and particles, such as liposomes and liquid crystals. This ability has made PC very useful as an excipient in pharmaceutical formulations. Other technically and biologically important double-chain phospholipids are phosphatidic acid (PA), phosphatidylethanol-

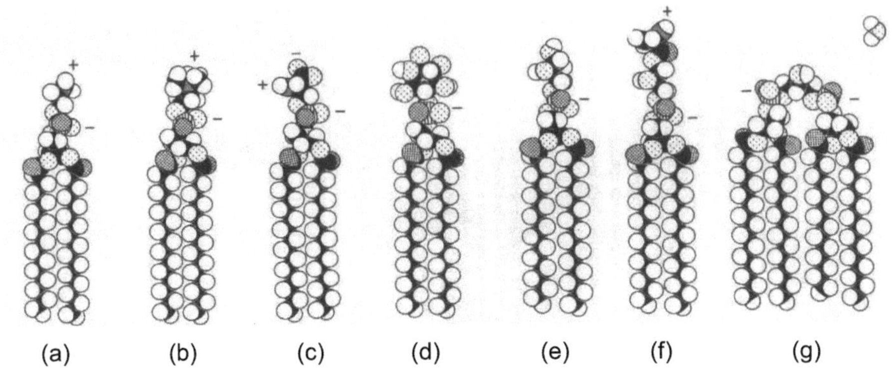

Figure 5.1. Space filling models of various phospholipids: (a) phosphatidylethanolamine (PE); (b) phosphatidylcholine (PC); (c) phosphatidylserine (PS); (d) phosphatidylinositol (PI); (e) phosphatidylglycerol (PG); (f) phosphatidyl-O-alanyl-glycerol; and (g) cardiolipin. The models reflect the atomic proportions of the non-polar acyl chains (hydrocarbon chains with 16 carbons) and the polar head groups, of which some are zwitterionic, others are anionic. Reprinted from Robertson (1983) with kind permission of Cambridge University Press.

amine (PE), phosphatidylglycerol (PG), phosphatidylinositol (PI) and phosphatidylserine (PS). Space filling models of these and some additional polar lipid molecules are depicted in Figure 5.1. The physical chemistry of phospholipids is indeed intriguing: their amphiphilic character results in diverse phase behaviour, particularly in water and aqueous solutions, but also in non-polar solvents. The amphiphilicity arises from the two non-polar fatty acyl chains and the polar head group attached to the glycerol backbone. Among the naturally occurring phospholipids, the head group may be electrically neutral but zwitterionic or negatively charged at neutral pH.

B. Basic properties

Some basic data of the most common double-chain phospholipids are listed in Tables 5.1–5.3. In Table 5.1, the molecular weights of the phosphatidylcholines include the weight of one water molecule; all the other values are for anhydrous lipids. For the long-chain 1,2-diacylphosphatidylcholines with saturated hydrocarbon chains (DMPC, DPPC, DSPC), the melting points vary relatively little with chain length (about 235°C) and the same holds for the 1,2-phosphatidylethanolamines (DPPE, DSPE). The alkyl derivatives (ether phospholipids) have rather similar melting points to the diacyl derivatives (data not shown). The melting points for the 1,2-diacylphosphatidic acids, 1,2-diacylphosphatidylglycerols and 1,2-diacylphosphatidylserines appear to increase somewhat with increasing chain length. The melting points of the acid forms of the

Table 5.1. Selected data for some common synthetic 1,2-diacyl-sn-glycero-3-phospholipids (PA, PE, PC, PG and PS)

Phospholipid	$R_1 = R_2$[a]	Common acronym	Molecular weight (g/mol)[b]	Melting point (°C)[c]	T_m (°C)[d]
PA					
1,2-Dilauroyl	12:0	DLPA	537.7		32
1,2-Dimyristoyl	14:0	DMPA	592.8	62	50
1,2-Dipalmitoyl	16:0	DPPA	648.9	70	68
1,2-Distearoyl	18:0	DSPA	705.0	76	66
1,2-Dioleoyl	18:1	DOPA	701.0		
PE					
1,2-Dilauroyl	12:0	DLPE	579.8	208	30.5
1,2-Dimyristoyl	14:0	DMPE	635.9	195	49.5
1,2-Dipalmitoyl	16:0	DPPE	692.0	186	65
1,2-Distearoyl	18:0	DSPE	748.0	180	74
1,2-Dioleoyl	18:1	DOPE	744.1	25	–16
PC					
1,2-Dilauroyl	12:0	DLPC	621.8	210	–1
1,2-Dimyristoyl	14:0	DMPC	677.9	237	23.5
1,2-Dipalmitoyl	16:0	DPPC	734.1	234	41.5
1,2-Distearoyl	18:0	DSPC	790.2	231	55.5
1,2-Dioleoyl	18:1	DOPC	786.1	227	–18
1,2-Dilinoleoyl	18:2		782.1	232	–53
1,2-Di-α-linolenoyl	18:3		778.1		–63
1,2-Diarachidoyl	20:0		846.3		66
1,2-Diarachidonoyl	20:4		830.2		
1,2-Dibehenoyl	22:0		902.4		75
1,2-Lignoceroyl	24:0		958.5		80
	R_1 R_2				
1-Palmitoyl-2-oleoyl	16:0 18:1	POPC	760.1		–3
PG					
1,2-Dilauroyl	12:0	DLPG	610.8		–2
1,2-Dimyristoyl	14:0	DMPG	666.9		24
1,2-Dipalmitoyl	16:0	DPPG	723.0		41.5
1,2-Distearoyl	18:0	DSPG	779.1	67	54.5
1,2-Dioleoyl	18:1	DOPG	775.1		
1,2-Dilinoleoyl	18:2		771.0		
	R_1 R_2				
1-Palmitoyl-2-oleoyl	16:0 18:1	POPG	749.0	176	
PS					
1,2-Dilauroyl	12:0	DLPS	623.8		13
1,2-Dimyristoyl	14:0	DMPS	679.9		36
1,2-Dipalmitoyl	16:0	DPPS	736.0	153	54
1,2-Distearoyl	18:0	DSPS	792.1	159	68

From Marsh (1990) and Cevc (1993).

[a] R describes the lipids' acyl chains: the first figure is the number of carbon atoms; the second is number of double bonds.

[b] Molecular weights for anhydrous phospholipids and for acid forms of PA, PG and PS (2 protons for PA; 1 for PG and PS).

[c] Melting points for hydrated forms of PC (1 H_2O per molecule); all other melting points for anhydrous phospholipids. If a melting point interval is reported, the lower value is shown without decimal.

[d] T_m is the phase transition temperature (gel to fluid/liquid crystalline), or the chain-melting temperature, in excess water.

Table 5.2. Logarithm of the binding constant (pK) for the stepwise protonation (decrease in pH) of the head group of double-chain phospholipids

Phospholipid	Amine group	Carboxyl group	Phosphate group
PA	–	–	6.2, 1.8
PE	9.3	–	1.9
PC	11.0	–	1.7
PS	9.6	3.0	1.8

From Tatulian, 1993.

Table 5.3. Apparent pK_a values of ionizable groups of some common double-chain phospholipids[a]

Phospholipid	Medium	Temperature (°C)[b]	pK_a (PO_4)	pK_a (COO^-)	pK_a (NH_3^+)
DLPA	0.1 M NaCl	T_t	4.0, 8.1	–	–
DMPA	0.1 M NaCl	T_t	4.0, 8.5	–	–
DLPE	Triton X-100	40	~1	–	9.8
DMPE	0.1 M NaCl	T_t	1.7	–	11.2
DMPC	0.1 M HCl	T_t	~1	–	–
DPPC	0.1 M HCl	T_t	~1	–	–
DMPG	0.1 M NaCl	T_t	3.5	–	–
DPPG	0.1 M NaCl	T_t	3.1	–	–
DMPS	0.1 M NaCl	T_t	2.6	5.5	11.6
DPPS	0.1 M NaCl	25	NA	4.6	NA

[a] The enthalpy of protonation is much higher for the amino group than for the phosphate and carboxylate groups and is thus more dependent on the temperature (Marsh, 1990).
[b] T_t is the temperature deduced at the gel-to-liquid phase transition from titration of the transition temperature.
NA = data not available; – = not applicable.

1,2-diacylphosphatidic acids, 1,2-diacylphosphatidylglycerols and 1,2-diacylphosphatidylserines are much lower than those of the zwitterionic phospholipids. The gel-to-liquid bilayer transition temperature denoted T_m or T_c, will be discussed in Section E.

Most phospholipids are soluble in chloroform, dichloromethane, hexane, methanol, ethanol, isopropanol, pyridine and in mixtures thereof. Commonly used in HPLC for the dissolution and elution of phospholipids are mixtures of hexane and isopropanol, sometimes with a small portion of water added. In general, this combination is preferred over most other solvents/solvent mixtures from a toxicological and an environmental perspective. Water and aqueous solutions are poor solvents for phospholipids; for example, the maximum solubility of double-chain phosphatidylcholines with 14–18 carbon atoms at room temperature is only about 10^{-10} M (Robertson, 1983; Cevc, 1993). However, phospholipids are readily dispersible in water, aqueous media and other polar solvents, a feature which will be detailed in later sections.

The relative density of anhydrous phospholipids (and galactolipids) is about 1.08 g/cm^3, as measured with the gas pycnometer method (Delogu Löwbeer, 1994).

Zwitterionic phospholipids contain both a negative and a positive charge under normal conditions. They are sometimes referred to as amphoteric substances, but since the term 'amphoteric' actually means that they can be cationic, zwitterionic or anionic depending on the pH (cf. an amino acid), and since PC retains one of its charges (the positive charge on the quaternary ammonium residue) over the whole pH range, phospholipids are not truly amphoteric.

Despite the low solubility, it is possible to attribute an apparent acid dissociation constant, that is, a pKa, to the different protonatable/deprotonatable groups in the phospholipid polar head group, as discussed by Marsh (1990) and Tatulian (1993). These groups consist of phosphate, amine and carboxyl moieties. Due to the surface effects of phospholipids, the pKa values of these groups are slightly different from the intrinsic values in bulk water presented in Table 5.2. Due to the sensitive experimental procedures employed, values also differ depending on the literature source. Nevertheless, it has been possible to determine apparent pKa values from the titration of the gel-to-liquid bilayer transition temperature at different bulk proton concentrations, and some pKa values are listed in Table 5.3 for the most common double-chain phospholipids (for a detailed review see Tatulian, 1993).

The protonation/deprotonation of the phospholipid head groups caused by a pH change in the aqueous environment has a great impact on the physical properties of the phospholipid molecule as a whole. A change in charge with pH (a change in sign and/or a change from zero charge to a net charge) may influence the molecular conformation, and thus the molecular packing parameter, and the overall polarity. In turn, this may affect, for example, phase transitions, hydration/swelling, and interactions with ions and solutes, such as peptides and proteins, and the colloidal stability of aqueous dispersions of phospholipid aggregates. When the solution pH equals the isoelectric point, the physicochemical behaviour of the phospholipids resembles that of non-ionic lipids such as mono- and digalactosyldiacylglycerols.

A change in pH may also influence the hydrolytic stability of the phospholipids, with more rapid hydrolysis at high pH than at low pH.

C. Amphiphilicity

All phospholipid molecules consist of two parts, one that is soluble in a particular liquid (the liophilic part) and another that is insoluble in the same liquid (the liophobic part) (Holmberg *et al.*, 2003). In the case of water, the parts are referred to as hydrophilic and hydrophobic, respectively. This is why surfactants and thus phospholipids also are termed amphiphilic molecules or amphiphiles, where the word 'amphi' is derived from Greek, meaning 'both'. The expression amphipatic is synonymous with amphiphilic. The hydrophilic part of the molecule is often called the head group, whereas the hydrophobic part is called the tail. In a micelle, the tails are thus directed towards the interior of the aggregate whereas the head groups are directed towards water (see Figure 5.2). The micelle may be regarded as a polar aggregate of high water solubility and low surface activity (since it is only the monomer which is affecting adsorption and reduction in surface tension). The hydrophobic part of a phospholipid (i.e., the acyl portion) may be single-chain or double-chain, linear without branches, and saturated or unsaturated to a certain degree. The unsaturation, if any, is normally 1–3 double bonds per acyl chain and has usually a great impact on the physical properties of the lipid, particularly in water. The polar part of a phospholipid in water or aqueous solution at normal pH may be anionic or zwitterionic (neutral with both positive and negative charges). A phospholipid with a negative net charge has significantly different physical properties in solution compared to a zwitterionic species. In particular, the interaction with water, or the hydration, is affected, resulting in much stronger swelling of a negatively charged phospholipid. Furthermore, a negative charge increases the colloidal stability of dispersed systems, such as emulsions and liposomes. However, in the presence of high salt concentrations, the charge may be effectively screened by the positively charged counter-ions that diminish the repulsive forces between different particles and eventually leads to phase separation.

Most anionic synthetic phospholipids are monovalent with sodium as the counter-ion, but other cations such as lithium, potassium and ammonium ions are possible. Even divalent cations are available, such as calcium and magnesium ions.

One advantage of zwitterionic phospholipids over the charged counterparts is that they possess excellent dermatological properties. From a formulation perspective, neutral and zwitterionic polar lipids are preferred since these function well at high electrolyte concentration, and do not interact unfavourably with other charged components such as polysaccharides and peptides. From a toxicological point of view, neutral and zwitterionic polar lipids are preferred since these are unable to interact electrostatically with components *in*

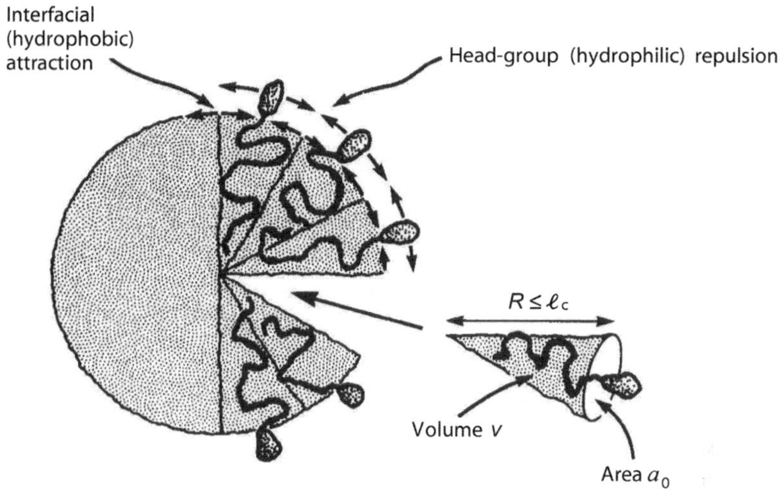

Figure 5.2. Schematic representation of a spherical micelle, with its polar head groups oriented towards water and the acyl chains in the centre in a liquid-like state. The critical packing parameter (CPP) relates the head group area, the extended length and the volume of the acyl chains into a dimensionless number CPP = $v/(l_{max}a)$. Reprinted from Israelachvili (1985) with kind permission from Elsevier.

vivo. However, lysophospholipids always exert their solubilizing or lytic capability *in vivo*, whether or not the head group is charged.

In addition to phospholipids, several other polar lipids are found in nature such as mono- and diacylglycerols, bile salts and glycolipids. Free fatty acids and their salts (soaps) are other examples. They are abundant in all living organisms, where they have many functions such as emulsifiers of fats and other hydrophobic substances in the gut and as building blocks of organized structures such as cell membranes. Digalactosyldiacylglycerol, a non-ionic glycolipid which can be isolated from many plant membranes, has many features in common with PC and other bilayer-forming phospholipids, including a strong swelling in water and the ability to stabilize emulsions and to form liposomes (Carlsson *et al*., 1996).

Phospholipids are examples of surface-active agents, or surfactants. Fundamental to surfactants is their tendency to adsorb and enrich at various surfaces and interfaces: that is, boundaries between any two immiscible phases (Holmberg *et al*., 2003). Liquid–vapour interfaces, where the vapour usually is air (present in foams, for example), liquid–liquid interfaces (emulsions), and liquid–solid interfaces (suspensions) are all examples of interfaces the formulator often encounters, sometimes all types in the same system, and where phospholipids may have a significant action. The driving force for a surface-active phospholipid to adsorb at a particular interface is to lower the free energy of that phase boundary. The term interfacial tension is frequently used instead of interfacial free energy per unit area. In the case of water, the most common liquid in most practical applications involving phospholipids, the surface tension is thus equivalent to the interfacial free energy per unit area needed to expand the water boundary exposed to air. When this boundary is covered with phospholipid molecules, the surface tension is reduced. A higher number of phospholipids at the interface (i.e., a denser packing) leads to a larger reduction in surface tension.

When an interface is fully covered with phospholipids or any other type of surfactant, the phospholipid monomers will start to form aggregates in aqueous solution, another fundamental property of surfactants. Single-chain surfactants form aggregates called micelles, normally spherical in shape with a finite aggregation number or number of monomers (monomers are also referred to as unimers) (Holmberg *et al.*, 2003). Micelle formation, or micellization, is another mechanism for a surfactant to prevent the hydrophobic portion coming into contact with water, which thereby reduces the free energy of the system. The concentration required for the onset of micellization is known as the critical micelle concentration (CMC). At higher surfactant concentration, the monomer concentration will not exceed the CMC, only the concentration of micelles will increase. Figure 5.3 depicts the changes that occur in micelle-forming systems. For example, the surface tension as well as the osmotic pressure becomes more or less constant, whereas turbidity and solubilization start to increase and self-diffusion starts to decrease. Thus, there is a transformation from a solution containing singly, molecularly dispersed polar lipid molecules or ions, to a situation where the polar lipids are more and more in an aggregated or self-assembled state; that is, spherical micelles (Lindman & Wennerström, 1980; Holmberg *et al.*, 2003).

Among phospholipids, only the single-chain classes are soluble enough and have a molecular geometry suitable for forming micelles (in general, double-chain surfactants including phospholipids do not form micelles). Lysophospholipids have not gained much attention in practical applications, but they have important physiological functions as, for example, natural emulsifiers of fatty substances in the gut and as absorption enhancers. CMC data for long-chain lysophospholipids are scarce, mainly because of their very low and not easily detected CMC; for example, CMC data for long-chain lyso-PC are reported to be of the order of 1 μM (Cevc, 1993). Nevertheless, some general remarks can be made on the variation of CMC with the chemical structure of a surfactant, i.e., of a lysophospholipid (Lindman & Wennerström, 1980; Holmberg *et al.*, 2003):

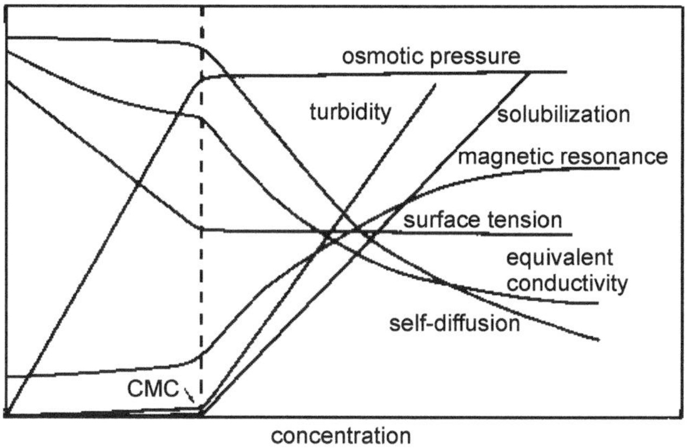

Figure 5.3. Schematic representation of the concentration dependence of some physical properties for solutions of a micelle-forming surfactant. Reprinted from Lindman and Wennerström (1980) with kind permission of Springer Science and Business Media.

i) The CMC decreases strongly with increasing acyl chain length; generally, it decreases by a factor of about three for zwitterionic surfactants on adding one methylene group to the acyl chain
ii) The CMC for a non-ionic surfactant is much lower than that of an ionic (at the same acyl chain length)
iii) For a given ionic surfactant any monovalent inorganic counter-ion gives roughly the same CMC, but a change to a divalent counter-ion gives a reduction by roughly a factor of four; an organic counter-ion results in a lower CMC compared to an inorganic one
iv) The introduction of branches and double bonds in the acyl chain has an affect on the CMC; a branched methyl group may cause an increase in CMC while one or more double bonds may cause a decrease
v) The CMC is more or less independent of temperature
vi) The CMC for an ionic surfactant is dramatically lowered in the presence of added salt, whereas that for a non-ionic surfactant is much less dependent.

D. Classification of phospholipids

One way to classify the hydrophilicity of a substance depends on how it is partitioned between two immiscible phases, normally octanol and water. A low value of the partitioning coefficient P, usually reported as log P, stands for a high hydrophilicity. Another way of classifying the hydrophilicity, particularly of non-ionic surfactants, is the concept of hydrophilic–lipophilic balance

(HLB). The relationship between log P and HLB is reverse: the higher the HLB value, the lower the log P value. HLB values are often used in connection with the performance of emulsifiers, particularly non-ionic ones of the ethylene oxide type (Bergenståhl, 1997). An HLB value of an unknown emulsifier may thus be determined by comparing the emulsification properties in a predetermined system of a mixture of hydrophobic and hydrophilic emulsifiers with a predefined, arbitrarily chosen HLB value. It is also possible to estimate the HLB value of an unknown emulsifier from the molecular formula using a group contribution system; in addition, large tables of data for commercial emulsifiers are available. However, it is difficult to calculate HLB values for phospholipids (Bergenståhl, 1997). Nevertheless, by comparing the emulsification properties with existing data for commercial emulsifiers, it is possible to assign HLB values to phospholipids. In Table 5.4, a comparison is made between various types of phospholipids. Normally, PC has intermediate HLB values in the range of 8–10 that makes this lipid dispersible in both polar and non-polar solvents such as water and triacylglycerol oils, respectively. Consequently, PC can emulsify and stabilize both water and oil-continuous emulsions, depending on the exact HLB value which in turn depends on the composition (purity) with respect to phospholipid classes. PE has a lower HLB value at neutral pH and suits even better as an emulsifier of oil-continuous emulsions. Lysophospholipids give rise to HLB values above 13 and are dispersible or even soluble in polar solvents such as water, and are thus useful for making water-continuous emulsions. Of course, all phospholipids may be combined and mixed in appropriate proportions in order to find the optimal emulsification properties for a given application. This is always done empirically, as it is difficult to predict the behaviour of complex mixtures of phospholipids, particularly if they are naturally derived. To summarize, a low log P value of a phospholipid corresponds to a high HLB, and such a lipid is useful for making oil-in-water emulsions; vice versa, a phospholipid with a high log P and thus a low HLB value is utilized in water-in-oil emulsions.

A second method of characterization is to calculate the molecular dimensions of the phospholipids and then consider the geometrical constraints when packing the molecules into various aggregates. This is the concept of critical packing parameter (CPP) introduced by Israelachvili (1985). Figures 5.2 and 5.4 show the definition of CPP and how this number is related to various aggregate structures, respectively.

Finally, according to Small (1986), polar lipids may be divided into three different classes depending on their interaction with water. These classes are the insoluble, non-swelling lipids (Class I); the insoluble and swelling lipids (Class II); and the soluble lipids (Class III). According to this classification,

Table 5.4. Some properties of phospholipids related to partitioning, hydrophilicity, molecular shape and appearance in aqueous solution

Phospholipid	**Log P**	**HLB**	**CPP**	**Molecular shape**	**Type of aggregate**	**Appearance in water**
LPC, LPG	Low	>13	< ⅓	Cone	Spherical micelle	Clear solution
PA, PC, PG, PI, PS, PE at low or high pH	Intermediate	8–13	½–1	Truncated cone	Flexible bilayer	Milky to translucent dispersion
Saturated PE	High	6–8	~1	Cylinder	Planar bilayer	Milky dispersion after agitation
Unsaturated PE	High	3–6	>1	Inverted truncated cone	Inverted micelle	Poor dispersibility

The HLB values are rough estimates and have been assigned by means of the appearance in water (according to Holmberg *et al.*, 2003). The properties of the phospholipids also depend on the acyl chain length and counter-ion of the phospholipid, as well as the solution properties (temperature, pH, ionic strength).

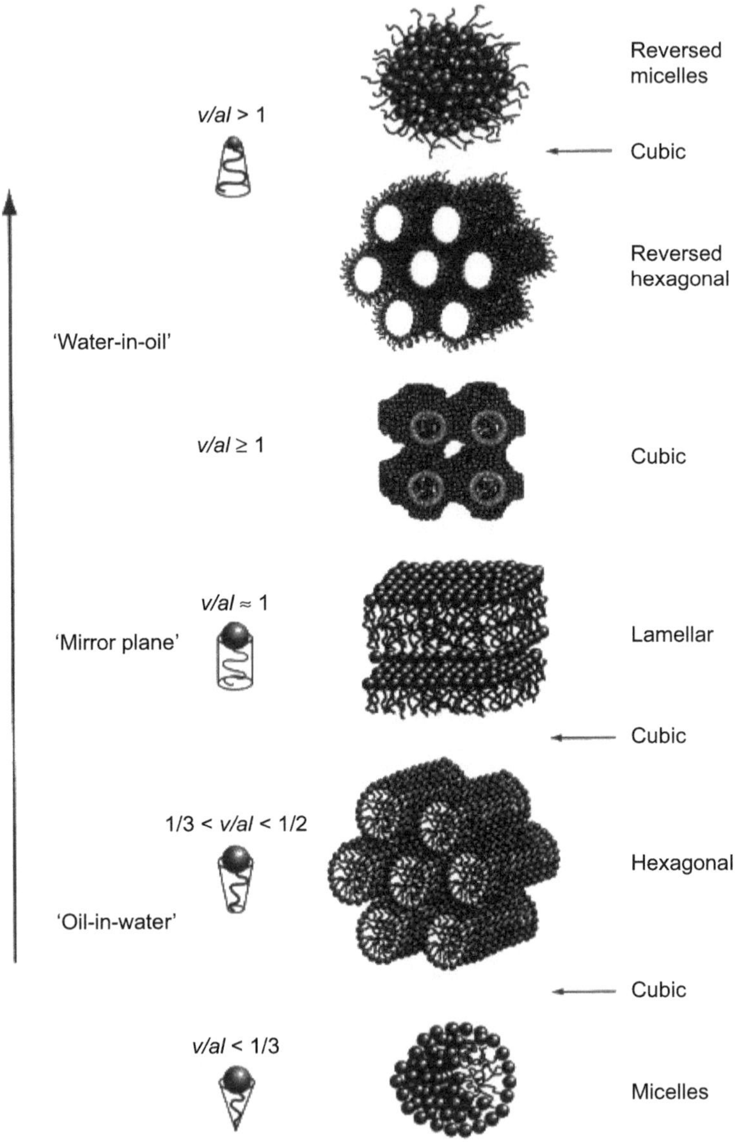

Figure 5.4. Critical packing parameters (CPP) of surfactant molecules and preferred aggregate structures for geometrical packing reasons. The arrow on the left illustrates the symmetry in curvature, from positive to negative, of the aggregates around the lamellar phase. The change in symmetry along the direction of the arrow can be effectuated by an increase in lipid concentration and/or an increase in temperature. Reprinted from Holmberg *et al.* (2003) with kind permission of John Wiley & Sons Limited.

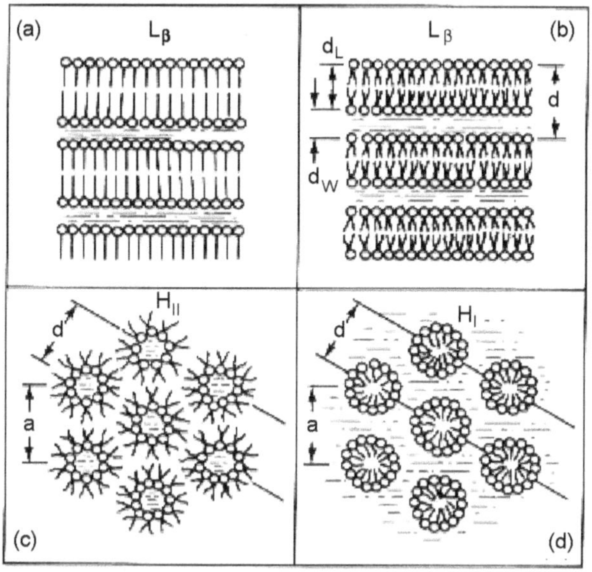

Figure 5.5. Schematic representation of some important lipid–water phases: (a) lamellar gel; (b) lamellar liquid crystalline; (c) reversed hexagonal; (d) hexagonal. The lamellar phase may be regarded as the mirror plane, where the aggregates are of the 'oil-in-water' type on the water-rich side and of the 'water-in-oil' type on the water-poor side (Fontell, 1992). For double-chain phospholipids, the normal hexagonal phase is not possible due to packing constraints (see Section D). Various dimensions that can be measured by X-ray diffraction are indicated. Reprinted from Gennis (1989) with kind permission of Springer Science and Business Media.

phospholipids belong either to Class II (the double-chain phospholipids) or Class III (the single-chain phospholipids). The Class II lipids are particularly interesting from a phase behavioural perspective; as previously mentioned, they are virtually insoluble in water, but still the interaction with water, or swelling in water, in a temperature-dependent fashion, results in the formation of various liquid crystalline phases. Among the Class I lipids are triacylglycerols, diacylglycerols and long-chain protonated fatty acids. Due to their negligible solubility in water in combination with a small but still effective polarity, these molecules can spread on water to form a stable monolayer (see Figures 5.4 and 5.5 for a summary of the various liquid crystal forms and Table 5.4 for the notations).

E. Various lipid phases and the chain-melting temperature

In general, the aggregation of polar lipids leads to a range of different structures, of which a few are shown in Figure 5.4. Systems containing a polar lipid may be divided into homogeneous (single-phase or one-phase) systems and heterogeneous systems (Holmberg *et al.*, 2003). The single-phase systems may further be divided into isotropic solutions, solid phases and liquid crystalline phases. The solid crystalline phases have both long-range and short-range order, with a varying degree between different phases (see above). Isotropic solution phases are disordered over both short and long distances, whereas liquid crystalline phases have a short-range disorder but some distinct order over larger distances. A high order over long distances in a liquid crystalline phase, induced by a drop in temperature or increase in concentration of the polar lipid, for example, may be manifested as a dramatic increase in the macroscopic viscosity, while the fluidity on the molecular level (the 'microviscosity') is still low. In both isotropic solutions and liquid crystals, the acyl chains are in a 'liquid-like' state. In solid crystals, the chains are in a more 'solid-like' state. The temperature at which the chains are transformed (or 'melted') to the 'liquid-like' or 'fluid-like' state is commonly referred to as the chain-melting temperature (T_m).

Chain-melting temperatures for some synthetic phospholipids in excess water are presented in Table 5.1. The following qualitative remarks can be made regarding the chain-melting temperature (Gennis, 1989):

i) T_m is most dependent on the length and the unsaturation of the acyl chains; longer chains result in higher T_m, because of the increased van der Waals attraction for the longer chains, whereas a double bond results in a dramatic decrease in T_m due to interruption of the optimal interactions between the chains in the gel state (more so for a *cis* double bond than for a *trans*);
ii) Saturated PA or PE have a T_m which is about 25°C higher than the corresponding PC;
iii) PC and PG exhibit similar phase transition temperatures for a given chain length, whereas those for PS are between PC/PG and PA/PE;
iv) Natural phospholipids normally have chain-melting temperatures below 0°C, but can be transformed to high-melting phospholipids by hydrogenation. The resulting T_m is then normally in the range of 45–50°C, depending on the original acyl chain composition in the natural starting material.

The chain-melting, that is, the phase transition of the acyl chains from a solid-like to liquid-like state, is connected with a positive enthalpy change, normally in the range of 25–45 kJ/mol. The corresponding entropy change is also positive, in the range of 100–130 J/mol/K (Marsh, 1990; Cevc, 1993).

The liquid crystalline state, sometimes referred to as the 'mesomorphic state', which indicates an intermediate state of matter, may be induced by changes in temperature, so-called thermotropic mesomorphism, and/or by the addition of a second component such as a solvent like water or other changes in composition, so-called lyotropic mesomorphism (Small, 1986). Furthermore, as seen in Figures 5.4 and 5.5, a liquid crystal may appear in many different forms. From a molecular perspective, the liquid crystalline state lies between the solid, crystalline state of matter and the fluid or liquid state. In general, liquid crystals are thus characterized by having long-range order and short-range disorder. The long-range order is manifested in high viscosity, birefringence and non-transparency of the liquid crystalline phase. Generally, the viscosity increases from the lamellar phase via the (normal or reversed) hexagonal phase to any of the cubic phases, which reflects the increase in long-range order; from order in one dimension (lamellar), in two dimensions (hexagonal), and in three dimensions (cubic). If a phase has order in one or more dimensions, it is anisotropic: it has different physical properties in different directions. Examples of physical properties that are dependent on the anisotropy are birefringence of polarized light, conductivity, diffusion (of lipid, water, and possible counter-ion and/or added solute), and nuclear magnetic resonance.

F. Phase behaviour in aqueous systems

Polar lipids are normally used in combination with water or aqueous solutions. The term 'polymorphism' is traditionally used to indicate that more than one (liquid) crystalline phase is formed by a substance. The polymorphic behaviour of polar lipids in water is thus technically as well as theoretically interesting. Polar lipids in general show a similar phase (polymorphic) behaviour.

Phospholipids, especially PC, are the polar lipids that have been the subject of most studies concerning aqueous interaction and phase properties, because of their importance in biological membrane functions, as well in technical applications (Larsson, 1994). This is illustrated by numerous publications in the field, recently reviewed and indexed by Koynova and Caffrey (2002) who present more than 2000 phase diagrams of various lipids, including phospholipids. The majority of the phase diagrams presented refer to binary (two-component, i.e., phospholipid–water) temperature-composition phase diagrams, but also more complex (ternary, quaternary) phase diagrams are listed. Also included in the index are phase diagrams constructed without water (dry systems) and a few with non-aqueous solvents such as glycerol. Thus, the index provides a convenient listing of lipid miscibility and phase behaviour, with complete bibliographic reference to the source material. The origin of the

Table 5.5. Different systems of notation for the most common liquid crystalline and other phases

Phase structure	Notations
Lamellar	lam, L_{α}, D, G, neat
Hexagonal	hex, H_{I}, E, M_1, middle
Reversed hexagonal	rev hex, H_{II}, F, M_2
Cubic (normal micellar)	cub (cub_m), I_1, S_{1c}
Cubic (reversed micellar)	cub (cub_m), I_2
Cubic (normal bicontinuous)	cub (cub_b), I_1, V_1
Cubic (reversed bicontinuous)	cub (cub_b), I_2, V_2
Gel	gel, L_{β}
Micellar	mic, L_1
Reversed micellar	rev mic, L_2
Microemulsion	μem, L, S, μE
Vesicular	ves

Adapted from Holmberg *et al.* (2003) with kind permission of John Wiley & Sons Limited.

index is the LIPIDAG database (http://www.lipidag.ul.ie), an entity of the web-based databases related to lipids and membranes formerly known as the Lipid Data Bank. The other entities are LIPIDAT, a compilation of lipid phase types and phase transition characteristics, and LMSD, a database of lipid molecular structures.

The structures of liquid crystalline phases encountered in phospholipid–water systems are depicted in Figure 5.4. The most common are the lamellar, gel and reversed hexagonal phases, which are presented in more detail in Figure 5.5. The various cubic phases exist only to a limited degree. Table 5.5 lists the various notations of the phases; various ways of notation are used by various authors, which may sometimes cause confusion.

The lamellar gel phase (or simply the gel phase) (L_{β}) has an intermediate structure between the (hydrated) solid state and the lamellar liquid crystalline phase. It is formed when water starts to interact with the polar head groups of the crystalline lipid molecules (see also Figure 5.6).

The lamellar liquid crystalline phase (L_{α}) represents the bulk of the lipids in a biological membrane. It is built up of bilayers of polar lipid molecules alternating with water layers. It is formed when the temperature is raised in a lamellar gel phase ($T>T_m$). The thickness of the bilayers is slightly less than twice the length of the lipid molecule because of the fluid state of the acyl chains (see also (a) and (b) in Figure 5.5). The phase has two-dimensional order, but there is considerable disorder in the acyl chains (Gennis, 1989; Holmberg *et al.*, 2003).

As already discussed, the phase transition temperature going from the gel

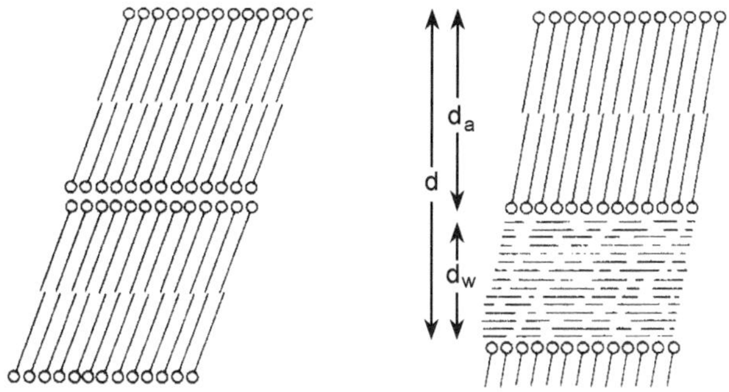

Figure 5.6. Formation of the gel state when water penetrates the gap between adjacent sheets of polar head groups of a polar lipid crystal structure. The polar head groups are indicated by circles and the axes of the acyl chains by straight lines. Reprinted from Larsson (1994) with permission of The Oily Press.

phase to the lamellar liquid crystalline phase is the chain-melting temperature. Normally, a 'low-melting' phospholipid means that the transition temperature of the acyl chains in excess water is low, for example, below room temperature or below the freezing point, not that the melting point of the anhydrous phospholipid as such is particularly low (cf. Table 5.1).

1. Phosphatidylcholine (PC)

In general, PC has a suitable hydrophilic–lipophilic balance to use for making artificial biomembranes, as the principal building block in liposomes and as an emulsifier in many food and pharmaceutical applications. PC from egg yolk is probably the most studied natural polar lipid (Larsson, 1994). The phase diagram in water is presented in Figure 5.7. At low water content, the phase behaviour is very complex and has been omitted (cf. the dashed area in this and the following figures) as it is less relevant in most applications. The lamellar L_α-phase dominates the phase diagram. At low water content, it is possible to obtain the transitions $L_\alpha \rightarrow$ cubic (C) $\rightarrow$ reversed hexagonal (H_{II}) by an increase in temperature. These transitions can be explained by an increase in volume of the acyl chains as the thermal mobility increases; that is, there is an increase in the critical packing parameter (cf. Figure 5.4). Although PC from egg yolk also contains minor amounts of other phospholipid classes, it behaves as one component. The maximal swelling for this particular material is about 44% at 25°C; above this water content, the system becomes immiscible and separates into two phases, L_α + water. The borderline between the L_α-phase and hydrated solid crystals and water, solid + H_2O, roughly corresponds to the chain-melting transition T_m, which is dependent on the water concentration up

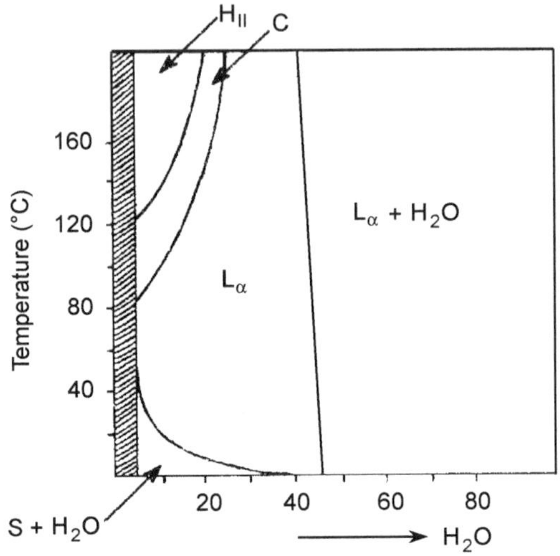

Figure 5.7. Schematic phase diagram of PC from egg yolk and water. L_α= lamellar phase; H_{II} = reversed hexagonal phase; S = hydrated solid crystals. Reprinted from Larsson (1994) with permission of The Oily Press.

to the maximal swelling limit. Above this limit, the chain-melting transition is below 0°C. Other PC mixtures of natural origin, such as PC from soybean, rapeseed and oats, have almost identical phase behaviour in water. It should be pointed out that swelling is increased already at very low content of charged components (see below), and may change from one source to another and even from one batch to another. *

DPPC is probably the most studied synthetic phospholipid. The phase diagram of DPPC in water is illustrated in Figure 5.8. Apart from a higher chain-melting temperature, the phase behaviour is quite similar to that of PC from egg yolk or other natural sources, with a dominating lamellar phase and the characteristic temperature-induced transitions from lamellar via cubic to reversed hexagonal phase at low water content. The chain-melting temperature, about 41°C at high water content, is almost constant above the maximal swelling limit (>36%).

* The hygroscopic properties and the heat of hydration of PC may be strong enough to cause accidents. The author has experienced cases where freeze-dried PC from soybean was taken out from the freezer without allowing the lipid material to attain room temperature in a desiccator. This generated an enormous amount of heat which caused melting of both the phospholipid material and the plastic container in which it was kept. A plausible explanation for this unexpected behaviour was the rapid exothermic uptake of atmospheric moisture, which generated a temperature increase which promoted the oxidation of the unsaturated acyl chains of the lipid material. The additional heat was more than enough to melt the plastic container and could have started a fire if not cooled down (cf. linseed oil on a ball of cotton waste in open air).

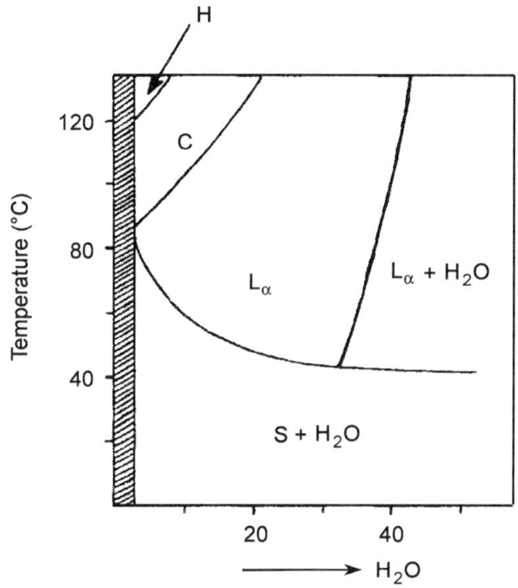

Figure 5.8. Schematic phase diagram of DPPC and water. Notations as in Figure 5.7. Reprinted from Larsson (1994) with permission of The Oily Press.

As previously pointed out, crystals of bilayer-forming lipids can form aqueous phases by penetration of water into the polar region of the bilayer, resulting in swelling of the crystals without changing the crystalline structure of the bilayer (see Figure 5.6). These phases are the gel phases and represent an intermediate structure between the solid state and the lamellar liquid crystalline phase (not indicated in the phase diagrams in Figures 5.7 and 5.8). When the L_α-phase of the DPPC-water system (or other systems containing saturated PC) is cooled, it is transformed into a gel phase. DPPC also exhibits a pre-transition state, with a rippled bilayer structure, at a water content of 20–30% and before the aqueous gel phase is transformed to the lamellar phase (Larsson, 1994).

The transition temperature from the gel phase to the lamellar phase is lowered from 55.5°C to 41.5°C when the acyl chain length is reduced from distearoyl to dipalmitoyl (DSPC → DPPC) and to about 23°C when further reducing the length to dimyristoyl (→ DMPC). For DLPC and DOPC, the corresponding transitions occur below 0°C (Larsson, 1994; see also Tables 5.1 and 5.6). The introduction of one double bond (in each acyl chain) in DSPC to form DOPC has the same effect as to shorten the acyl chain length by more than six carbons. Also 1-palmitoyl-2-oleyl-PC (POPC), a phospholipid with asymmetric acyl chains, has a chain-melting temperature of –3°C (Table 5.1). This points to the great impact of a double bond on the fluidity and the melting/transition point depression of the acyl chains. To illustrate this further, one may compare DSPC with the fully saturated, asymmetric PC containing one palmitoyl chain and one stearoyl chain (PSPC) which has a transition temperature of 49°C (Cevc, 1993), close to the arithmetical mean of DPPC (41.5°C) and DSPC (55.5°C), and with POPC. The elimination of two methylene groups in one of the acyl chains (R_1) reduces the transition temperature by only 6.5°C, whereas further elimination of two hydrogen atoms in the second acyl chain (R_2) reduces the transition temperature by more than 55°C.

A PC with one palmitoyl chain and one shorter chain behaves in a similar way to the previously described phosphatidylcholines, with respect to the formation of a dominating lamellar phase, down to a chain length of eight carbons (1-octanoyl-2-palmitoyl-PC). A further reduction in chain length to six carbons (1-hexanoyl-2-palmitoyl-PC) results in a phase behaviour that resembles more the behaviour of lysophospholipids (Larsson, 1994).

In this context, it is interesting to note that sphingomyelin (see Chapter 8), with a long saturated acyl chain, is a highly asymmetric polar lipid which can form an interdigitated bilayer. In an interdigitated bilayer, the long saturated acyl chains in one monolayer cross the bilayer centre and appose the shorter sphingosine chains from the other monolayer (McIntosh *et al.*, 1992). This results in increased bilayer stability through attractive van der Waals forces, further stabilized by intra- and intermolecular hydrogen bonding. The interdigitation is reduced by the addition of cholesterol, which can interact favourably with the long and saturated acyl chain (McIntosh *et al.*, 1992). Synthetic phospholipids with a large chain length difference also exhibit interdigitation, especially at low temperatures (Cevc, 1993; Chen *et al.*, 2001).

2. *Phosphatidylethanolamine (PE) and mixtures of PE and PC*

PE is an example of a major double-chain phospholipid that forms a 'non-lamellar' structure, the reversed hexagonal phase (H_{II}), in water, to an extent that strongly depends on the saturation of the acyl chains at a given chain length and the temperature. This is exemplified by DPPE in Figure 5.9 and mixtures of DOPE and DOPC in Figure 5.10. DPPE has a smaller and less hydrophilic head group than DPPC, which have obvious consequences on the phase behaviour in water (Seddon & Cevc, 1993). The chain transition temperature is higher, the region of the lamellar phase becomes smaller and the region of the reversed hexagonal phase becomes larger compared to DPPC (cf. Figures 5.8 and 5.9). The extension of the reversed hexagonal phase becomes even larger for DOPE, which forms a reversed hexagonal phase with 40% water at room

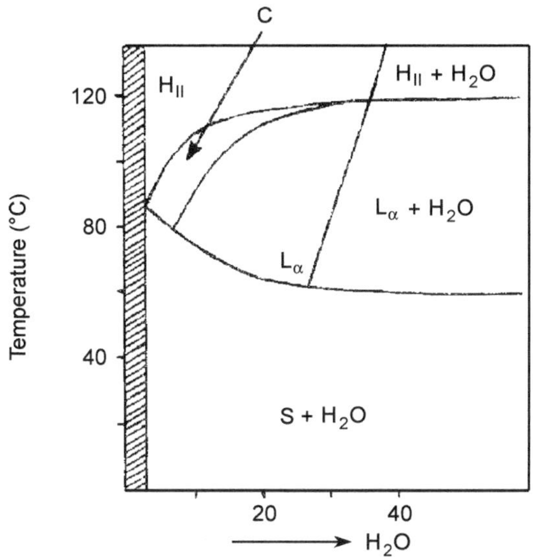

Figure 5.9. Schematic phase diagram of DPPE in water. Reprinted from Larsson (1994) with permission of The Oily Press.

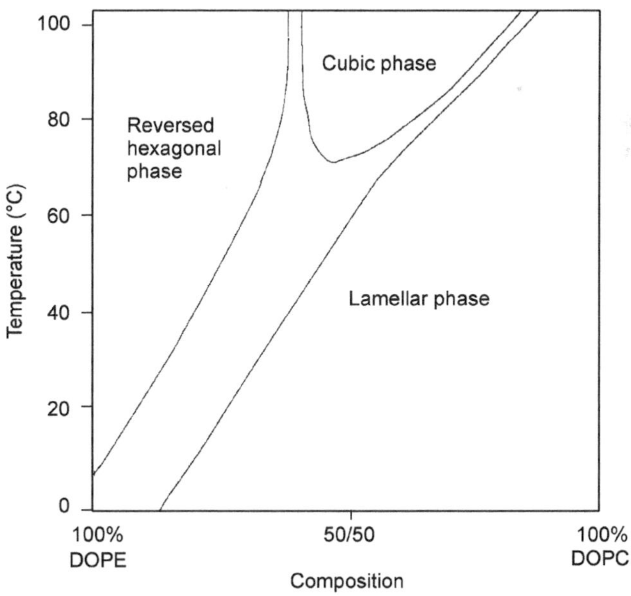

Figure 5.10. Schematic phase diagram of mixtures of DOPE and DOPC in 40% water. Reprinted from Bergenståhl (1997) with kind permission of Springer Science and Business Media.

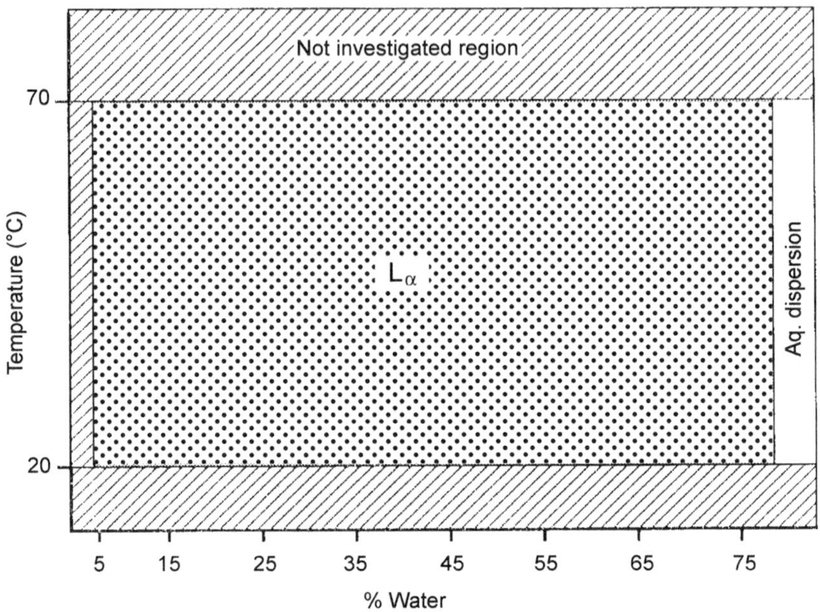

Figure 5.11. Schematic phase diagram of the sodium salt of PI from soybean in water. Reprinted from Söderberg (1990).

temperature (Figure 5.10). When mixing DOPE with DOPC in water the two components act like one, resulting in a balanced behaviour and the formation of lamellar phase above about 25% DOPC at 20°C (Figure 5.10). Like PC, PE is zwitterionic at neutral pH, but, unlike PC, becomes negatively charged at high pH and positively charged at low pH. In the ionic state, PE behaves in a similar way to the ionic phospholipids discussed below.

3. Anionic phospholipids

Phosphatidic acid (PA), phosphatidylglycerol (PG), phosphatidylinositol (PI), and phosphatidylserine (PS) are all negatively charged at neutral pH, and have a similar phase behaviour characterized by an extended lamellar phase with strong swelling and large separation between the bilayers. This phase behaviour is represented by the sodium salt of PI in Figure 5.11, which has a swelling in water up to about 77% (Söderberg, 1990). Above the swelling limit, a two-phase system (dispersion) is formed. If the monovalent counter-ion sodium is replaced by the divalent calcium ion, the repulsive forces between the polar head groups are diminished and a reversed hexagonal phase is formed. The same occurs if the charge is neutralized by a change in pH.

Presented in Table 5.6 is a summary of the type of liquid crystals formed, the chain-melting transition and the swelling limit for some common phospholipids.

4. Mixtures of phospholipids

A single phospholipid material or class is seldom used alone in practical applications; often other components are included. The reason for this may be to alter the hydrophilic–lipophilic balance, to decrease the fluidity of a bilayer, or to improve the colloidal stability of a dispersion, and so on. The additional component may be another phospholipid or another type of lipid with different polarity.

The effect of mixing PE and PC on the phase behaviour has already been discussed in relation to Figure 5.10. Ternary mixtures of PE–PC–water, PE–PI–water and PI–PC–water at a fixed temperature are shown in Figure 5.12 (Bergenståhl, 1997). Among the three phospholipids, the polarity decreases in the order: PI>PC>PE. This is clearly reflected in the ternary phase diagrams in Figure 5.12, where the lamellar liquid crystalline phase dominates in PI–water and PI–PC–water systems, and the reversed hexagonal phase in PE–water and PE–PC–water systems. When PE and PC are mixed in water, the properties of the mixture are dominated by PC up to about 25% PE (lamellar phase) and by PE (reversed hexagonal phase) above about 32% PE. When PE and PI are mixed, the properties are dominated by PE up to about 20% PI (reversed hexagonal phase) and by PI (lamellar phase) above about 21% PI (all percentages are taken at zero water content). Finally, when PI and PC are mixed, the properties are dominated by PI, seen as an increased maximal swelling at high water content.

A similar increase in swelling can be expected if PI is replaced by PG and/or PS, as long as they possess their negative charge. Phospholipids from soybean or any other natural source may thus be modified by the addition of additional phospholipid classes, particularly negatively charged, to balance the properties to fit a particular application. Of course, they may also be selected depending on the original composition; technical mixtures, originally containing high amounts of a charged species (PA, PG, PI, PS), are chosen for applications where it is preferable to have structures with a positive mean curvature. Similarly, mixtures rich in unsaturated PE are chosen when it is desirable to have structures with a negative mean curvature.

A mixture of two synthetic phospholipids may possess a chain-melting temperature which is in between the transition points of the individual components. For example, an equimolar mixture of DPPC and DSPC in excess water has an 'apparent' T_m of about 49°C, above which a single lamellar liquid

Table 5.6. The formation of liquid crystalline phases by various double-chain phospholipids

Phospholipid	**Major fatty acids**	**Type of liquid crystalline phases formed**	**Upper swelling limit (at 25°C)**
Phosphatidylcholines:			
DSPC	18:0	Lamellar phase at 55.5°C	–
DPPC	16:0	Lamellar phase at 41.5°C	36%
DMPC	14:0	Lamellar phase at 23.5°C	40%
DOPC	18:1	Lamellar phase below 0°C	42%
PC from egg yolk	16:0, 18:0, 18:1, 18:2, 20:4, 22:6	Lamellar phase below 0°C	44%
PC from soybean	16:0, 18:0, 18:1, 18:2, 18:3	Lamellar phase below 0°C	35%
Phosphatidylethanolamines:			
DPPE	16:0	Lamellar phase at 68°C, reversed hexagonal at 84°C	20%
DOPE	18:1	Lamellar phase below 0°C, reversed hexagonal at 5°C	20%
PE from soybean	16:0, 18:0, 18:1, 18:2, 18:3	Reversed hexagonal phase above 0°C	30%
Phosphatidylinositol:			
PI from soybean	16:0, 18:0, 18:1, 18:2, 18:3	Lamellar phase below 0°C	77%
Phosphatidic acid:			
DOPA	18:1	Lamellar phase below 0°C	High

Adapted from Bergenståhl (1997) with kind permission of Elsevier.

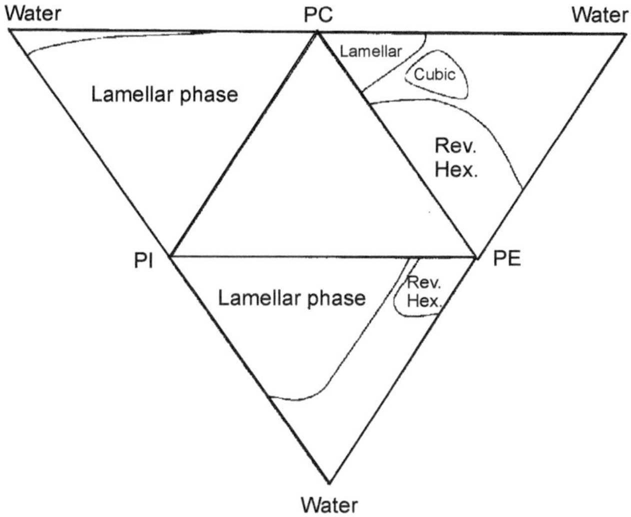

Figure 5.12. Schematic ternary phase diagrams of PC–PE and water; PC, PI and water; and PI (Na salt), PE, and water. The mixtures were investigated at room temperature. All three phospholipids were fractionated from soybean lecithin. Reprinted from Bergenståhl (1997) with kind permission of Springer Science and Business Media.

crystalline phase is formed. These two lipids mix ideally both in the gel state and the liquid crystalline state; that is, they are fully miscible, because of the small difference in acyl chain length. If the difference is larger, the pseudo-binary mixture may behave non-ideally, which has been reported for DLPC/DSPC and DMPC/DSPC mixtures (Gennis, 1989).

5. *Mixtures of phospholipids and cholesterol*

Thanks to the abundant occurrence of cholesterol in animal membranes, its interactions with phospholipids, particularly PC, have been investigated in detail. In aqueous phospholipid bilayers, the simple picture is that the hydroxyl group of cholesterol is situated at the interface between the acyl chains and water, whereas the planar and rigid ring structure and the hydrocarbon tail are located in the hydrophobic interior of the bilayer (for the cholesterol structure, see, for example, Gunstone & Herslöf, 2000). However, specific interactions give rise to unusual effects. This is exemplified in Figure 5.13, which shows the phase diagram for DPPC–cholesterol in excess water. Here, a notation is used that accounts for both two-dimensional translation order and chain conformational order occurring in the phases. Thus, the gel phase is referred to as the

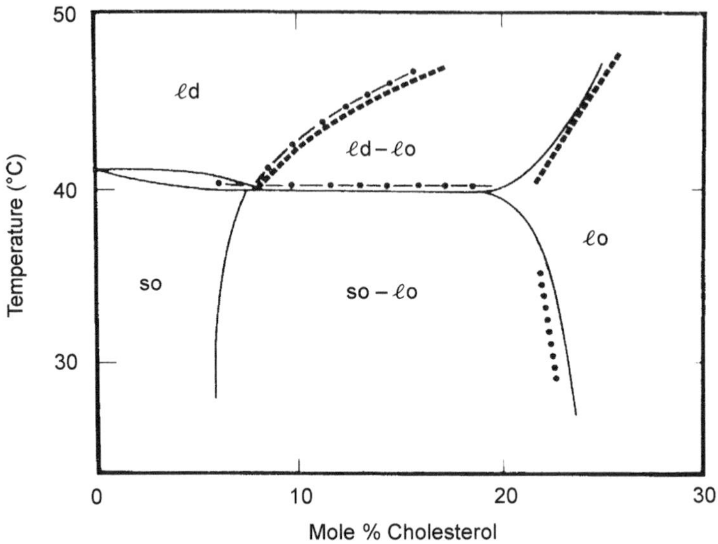

Figure 5.13. Experimental phase diagram for the system DPPC–cholesterol in excess water as determined by deuterium nuclear magnetic resonance spectroscopy and differential scanning spectroscopy (——), electron paramagnetic resonance (-------), freeze-fracture electron microscopy (••••••••), and micromechanics (• – • – • – •) by various investigators. Data have been scaled appropriately to facilitate the comparison within a single figure. See text for explanation of the notations. Reprinted from Hjort Ipsen *et al.* (1987) with kind permission of Elsevier.

solid-ordered (so) phase, where s, for solid, refers to the crystalline order and o, for ordered, to the average acyl chain conformation. Similarly, the lamellar liquid crystalline phase at low cholesterol content is denoted liquid-disordered (ld) whereas at high cholesterol content there is a liquid-ordered (lo) phase (Hjort Ipsen *et al.*, 1987).

At low concentrations of cholesterol, the chain-melting transition of DPPC is not affected by cholesterol (normally a minor amount of an 'impurity' would cause a freezing-point depression). At intermediate concentrations, about 10 to 20 mole%, two liquid phases of the bilayer coexist. However, the most remarkable feature is the effect at high cholesterol concentrations, above about 25 mole%. Now, cholesterol stabilizes a liquid-ordered (lo) phase in which the bilayer still is liquid crystalline, but where the acyl chains are in a more expanded and ordered conformation due to favourable interactions with cholesterol. This feature is also referred to as a 'condensation' effect of the acyl chains in the bilayer. Thus, while maintaining the liquid crystalline state of the bilayer, cholesterol also dramatically reduces the membrane permeability due to closer packing of the expanded acyl chains of the phospholipids. The effect is general: both unsaturated and saturated phospholipids exhibit condensation effects upon addition of cholesterol. All natural phospholipid membranes are in the beneficial liquid crystalline state, which of course prevents the lipids from crystallizing and precipitating, but also allows for high membrane elasticity and high lateral diffusion and thus rapid ‘self-healing’ if any defects in the membrane occur (see Section I).

The unusual features of the phase diagram for PC–cholesterol imply that cholesterol is a unique molecule whose function in membranes is unlikely to be mimicked by other biochemical substances (Hjort Ipsen *et al.*, 1987).

6. Mixtures of phospholipids and medium-chain monoacylglycerols

The properties of a bilayer based on PC may also be adjusted by the addition of another type of polar lipid. This is exemplified in Figure 5.14, which presents the pseudo-binary phase diagram for a system consisting of PC from soybean and a naturally derived medium-chain monoacylglycerol (MCM), containing approximately 80% octanoyl (8:0) chains and 20% decanoyl (10:0) chains (Bohlinder *et al.*, 1994; Carlsson *et al.*, 1994). The MCM used is a fractionated material and does not contain any traces of di- and triacylglycerols. In the investigated concentration range and at room temperature, the lipid mixture forms two layered structures: a solid, crystalline structure at low PC content and a lamellar liquid crystalline structure above approximately 50% PC. The two anisotropic phases are separated by a fluid isotropic solution phase, L_2, with a eutectic point at approximately –20°C. The low-viscous L_2-phase consists of mixed reversed micelles in which the hydrophilic parts of the lipids are in the centre and the lipophilic parts are oriented outwards in the micelles.

The ability of the PC–MCM–mixture to take up water is relatively high (mixtures such as this, containing polar lipids, normally pre-absorb a small amount of atmospheric water, and are thus pseudo-binary mixtures). The maximal uptake of water, or swelling, of pure PC from soybean and MCM is around 36% and 50%, respectively. This is indicated in Figure 5.15 along the two axes towards the water vertex on the left. The lipids are completely miscible even at low water content and form an L_α-phase independent of the lipid composition. The phase boundary between the two pure lipids is more or less linear. The appearance of the L_α-phase is the same within the whole one-phase region. The only macroscopic difference is the viscosity; a higher proportion of MCM in the mixture and/or higher proportion of water give rise to a lower viscosity. A similar phase diagram, with a dominating L_α-phase at a relatively constant swelling, is also reported for the aqueous system of

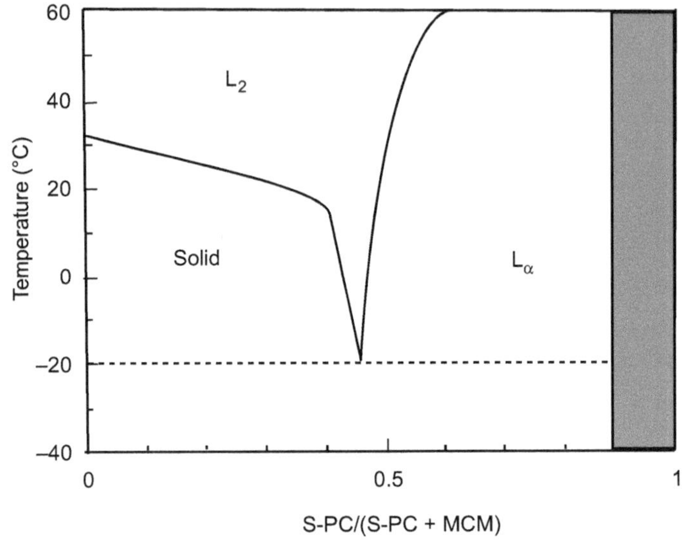

Figure 5.14. Schematic phase diagram for the binary system medium-chain monoacylglycerol (MCM) and phosphatidylcholine from soybean (S-PC) based on DSC data. The compositions are given as weight fraction of S-PC in the S-PC–MCM mixture. The symbols L_α and L_2 denote lamellar liquid crystalline phase and reversed micellar phase, respectively. For simplicity, the extensions of the two phase regions between the one-phase regions are not shown. Shaded area = region not investigated. Reprinted from Bohlinder *et al.* (1994) with kind permission of Elsevier.

1-monocaprylin (8:0) and DOPC (di-18:1) at 25°C (Söderberg, 1990). Since only one L_α-phase exists in equilibrium with water, it is clear that the 18:1-chains in DOPC and the corresponding ones in PC from soybean are completely miscible with the 8:0-chains in monocaprylin (and 10:0 in MCM), despite the difference in chain length in the liquid crystalline state (the chain transition temperature of pure MCM/monocaprylin is below 0°C in excess water). The packing of the two lipids in the bilayer does not result in constraints that would cause the lipids to segregate and to give, for example, two co-existing L_α-phases with different bilayer thickness.

In excess water, that is, above the swelling limit, large multilamellar vesicles start to form spontaneously (L_α + water in Figure 5.15). This may be assessed by the appearance of halos and Malthesian crosses in a light microscope with polarizers. The liposome formation is preceded by the growth of myelin figures from the lipid phase into the aqueous phase (Figure 5.16). Myelin figures are tubular, helical structures also found for some other swelling phospholipid systems, such as PC from egg yolk (Small, 1986) and hydrogenated PC from egg yolk mixed with cholesterol (Khan & Tucker, 1992).

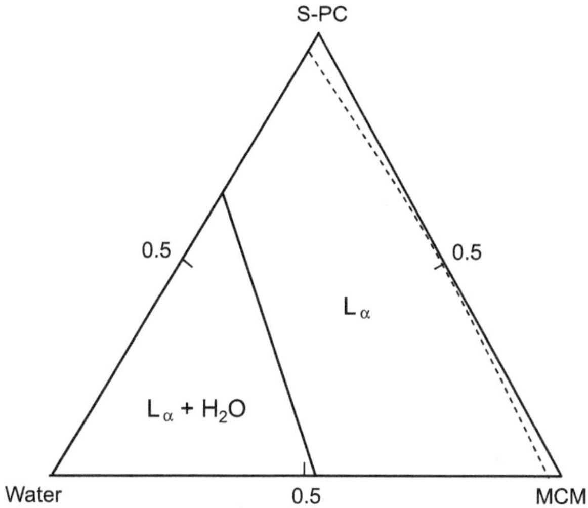

Figure 5.15. Schematic ternary phase diagram for the system consisting of PC from soybean (S-PC), MCM and water at 22°C. A characteristic feature of the system is the lamellar liquid crystalline phase, L_α, which dominates the phase diagram; i.e., the lipid mixture is well-balanced with respect to formation of a bilayer-forming structure. At low water content and at each lipid vertex, solid, crystalline phases appear (dashed lines). Reprinted from Bohlinder *et al.* (1994) with kind permission of Elsevier.

The rate of myelin formation and consequently the rate of liposome formation are dependent on the composition of the original lamellar phase. The rate is slow for a composition containing about 75% PC from soybean (thus allowing enough time for the picture shown in Figure 5.16 to be taken), whereas the rate is higher for mixtures with a high content of MCM. These macroscopic observations are supported by the data in Figure 5.17, where kinetics of hydration for various mixtures of PC from soybean and MCM with D_2O are presented (Carlsson *et al.*, 1994). Among the compositions studied, the measured hydration is rapid for the mixture containing 75% MCM and slowest for that containing 75% PC. Pure PC also hydrates rapidly. The kinetic data correlate with microcalorimetric data at 37°C in that the slowest hydrating compositions, in the range of about 50 to 90% PC, also show the highest heat of hydration (Carlsson *et al.*, 1994). In this concentration range and at 37°C, the lipid mixture forms an L_α-phase (see Figure 5.14).

The phase behaviour of the PC–MCM system in water is well balanced in that it forms a lamellar liquid crystalline phase more or less independently of the lipid composition. In excess water, this phase is unstable and forms liposomes at a rate that is dependent on the initial lipid composition. If monoolein is used instead, a long-chain, monounsaturated monoacylglycerol, more

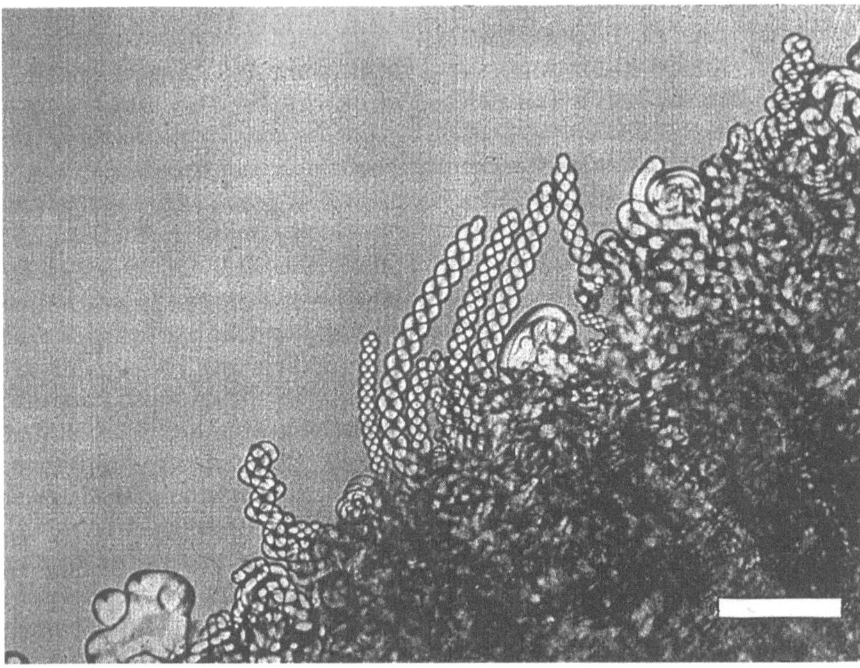

Figure 5.16. Myelin formation, as viewed in a light microscope, from a lipid mixture initially containing 74% PC from soybean and 26% MCM (by weight) in contact with water at room temperature. The photo is taken with partially crossed polarizers at a magnification of 100. The bar corresponds to 50 μm. Reprinted from Bohlinder *et al.* (1994) with kind permission of Elsevier.

complex phase behaviour in water is observed. The ternary system mono-olein–DOPC–heavy water has been investigated by NMR at 28°C (Gutman *et al.*, 1984). The isothermal phase diagram shows the existence of no less than four one-phase areas, containing homogeneous liquid crystal structures and one area with isotropic solution. Two of the liquid crystalline phases have a lamellar structure, one has a cubic, and the fourth a reversed hexagonal structure. Without water, a mixture of 3–5 moles of mono-olein per mole DOPC forms an oily liquid.

7. Lysophospholipids

Lysophosphatidylcholine (and free fatty acids) normally results from hydrolytic cleavage of PC, which may be caused by a change in pH or by the catalytic action of enzymes. Mixtures of mono- and diacylphospholipids are being used commercially as a combined solubilization aid and delivery device for poorly

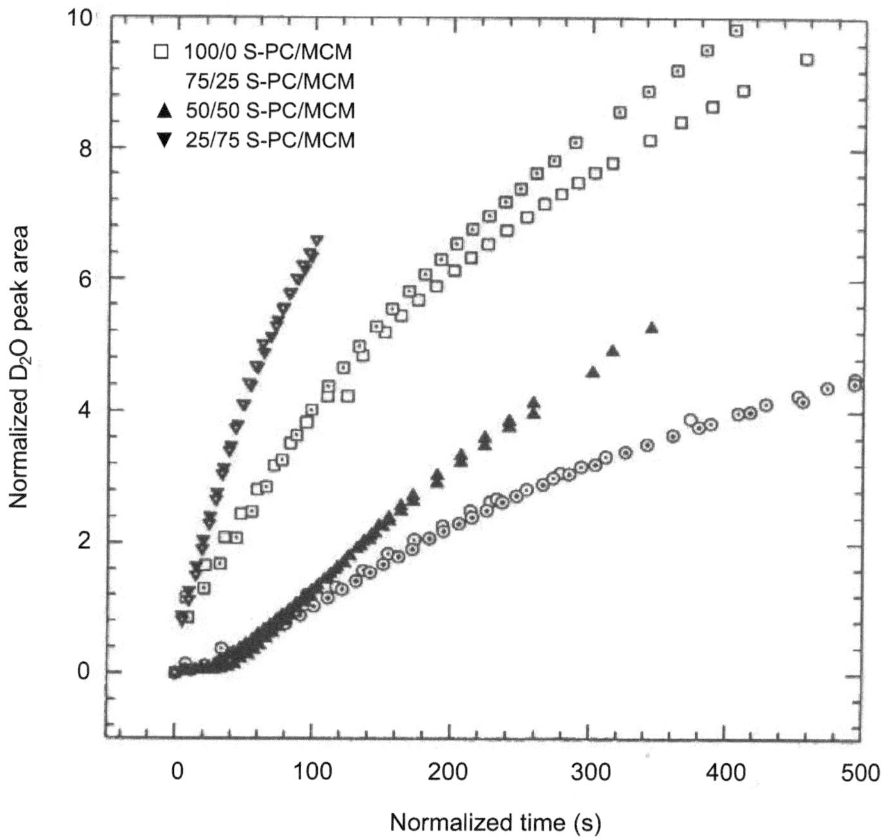

Figure 5.17. Kinetics of hydration of lipid mixtures measured by means of attenuated total reflectance (ATR) Fourier transform infrared spectroscopy. The sample was applied on an appropriate ATR crystal and initial spectra were recorded. D_2O was carefully layered on the sample which then was scanned repeatedly. Both D_2O peak areas, measured at 2488 cm^{-1}, and time were normalized. Measurements were made in duplicate. Reprinted from Carlsson *et al.* (1994) with kind permission of Biomedical Research Foundation.

water-soluble compounds (Leigh *et al.*, 2001). Such mixtures may be prepared by controlled cleavage of phospholipid blends by enzymes such as phospholipase A_1 and A_2 (Joshi *et al.*, 2006). Pure lysophospholipids have detergent-like properties in water and form micellar solutions at low concentrations and lamellar and normal hexagonal phases at high concentrations with cubic phases at intermediate concentrations (Arvidson *et al.*, 1985; Larsson, 1994). Generally, an increase in acyl chain length promotes the transformation from spherical to rod-like micelles.

Since all phospholipids are susceptible to hydrolysis, samples containing water have a limited shelf life. For example, aqueous PC forms lyso-PC and free fatty acids on standing. For a lamellar liquid crystalline phase based on PC and water, left at room temperature for several months, the hydrolysis is observed as a gradual increase in transparency, from turbid and birefringent to clear, and a gradual decrease in viscosity. This is due to the micelle-forming property of lyso-PC (and salts of free fatty acids at high pH) which gradually solubilizes the remaining double-chain PC in the mixture. At an intermediate stage, before all PC has been hydrolysed, the mixture may be regarded as a mixed micellar system. The problem of chemical instability may be overcome by replacing water with another polar solvent which is less apt to cause solvolysis, such as propylene glycol or glycerol (even if the latter solvent may cause interesterfication), but can still be swelled by the lipid.

For more information on phase diagrams of phospholipids in water in particular, and on phase behaviour of polar lipids in general, the reader is referred to, for example, Marsh (1990), Seddon and Cevc (1993), and Koynova and Caffrey (2002).

8. Experimental techniques for studying the phase behaviour of phospholipids

A great number of experimental techniques can be used to determine the physical properties of a phospholipid system, whether it is a micellar solution or a liquid crystal. Some of these are indicated in connection with Figure 5.13. Some of the most important techniques are polarizing microscopy, X-ray crystallography, NMR spectroscopy, rheology, surface balance and differential scanning calorimetry.

The easiest way to characterize a sample is to observe if it scatters normal light by ocular inspection or turbidity measurement using a spectrophotometer. Isotropic micellar solutions and cubic liquid crystals are clear and transparent, whereas the anisotropic liquid crystals scatter light and appear more or less cloudy (Holmberg *et al.*, 2003). Polarizing microscopy, perhaps the most affordable instrumental technique, can be used to judge if a sample is isotropic or not, since such a sample appears black between crossed polarizers, but also to distinguish between different anisotropic phases, such as lamellar and hexagonal phases.

Based on knowledge of the composition, it is often possible to guess the phase structure from the appearance in the microscope, but the structure must however be concluded by a more reliable technique. Traditionally, 'powder' X-ray crystallography has been used to accomplish this (Fontell, 1981), nowadays complemented by various NMR techniques, especially observations of quadrupole splitting in deuterium NMR (Lindman, 1987; Holmberg *et al.*, 2003).

Viscosity measurements may provide useful information on the rheological properties of phospholipid–water systems in a relatively simple manner. For micellar solutions, the viscosity is generally related to the micelle size and shape. The most important effect is the dramatic increase in viscosity occurring on formation of rod-like micelles, a process promoted, for example, by an increase in lysophospholipid concentration or by the addition of salt (for ionic lysophospholipids).

Generally, the viscosity of liquid crystals increases in the order lamellar < hexagonal < cubic. For the lamellar phase, the lipid bilayers can easily move over each other, but motion perpendicular to the layered structure is difficult. For hexagonal phases (both normal and reversed), motion along the cylinders is easy, but motion perpendicular to the cylinders is difficult since it involves changes of cylinder packing. Finally, for cubic phases there is no easy flow direction, since a three-dimensional structure exists and often a complex rheological behaviour is observed (shear thinning at low shear rates, increased viscosity at high shear rates, and so on).

Differential scanning calorimetry (DSC) is a common technique for the thermal analysis of pure and, ideally, anhydrous phospholipids, used for example for determination of the melting point, but more often for the determination of chain-melting temperatures and phase equilibria in binary phospholipid–water mixtures. For example, the schematic boundaries in the phase diagrams presented in this chapter may be more accurately measured using DSC. However, a complete phase diagram of a binary phospholipid–water system is difficult to construct, and great care should be taken regarding sample preparation and thermal history of the sample, particularly at high water content where dispersions are formed. A critical account on phase studies of phospholipid–water systems, particularly performed by calorimetry is given in a review by Chernik (1995).

The surface balance is an important tool for model studies of lipid monolayer properties at the water–air interface (Larsson, 1994). Pressure–area isotherms of a phospholipid on a pure water substrate at different temperatures may provide information that can be correlated to the chain melting in hydrated bilayers. Surface balance studies have been very useful for the understanding of the physical properties of artificial bilayers and liposomes, as well as natural cell membranes.

For a more detailed presentation of the whole range of techniques for studying the phase behaviour and other physical properties of phospholipids and other polar lipids, the reader is referred to, for example, Holmberg and co-workers (2003) and Gennis (1989).

G. Dispersed systems

Liposomes and emulsions are examples of colloidal dispersions in which highly subdivided phospholipid particles are surrounded by a continuous medium (polar or non-polar solvent). The liposomal dispersions may be normal or reversed, and the emulsions may be of oil-in-water or water-in-oil type. All dispersions are unstable from a thermodynamic perspective but possess kinetic stability. A typical particle size is in the range of about 20 nm to several μm. A small average particle size with a narrow distribution is preferable since this prevents or at least slows down the sedimentation on standing. Aggregation is prevented by electrostatic and/or steric repulsion. Sedimentation and aggregation may lead to coalescence and subsequent macroscopic phase separation.

In a water-continuous system, the colloidal stability may be increased by the addition of a water-soluble polymer. The polymer may increase the viscosity of the continuous phase and thereby decrease the sedimentation rate and self-diffusion. In addition the polymer may also adsorb on to the surface of the particles and thereby stabilize them through steric repulsion. Surface charges imposed by charged phospholipids, for example, prevent the oil droplets from fusing and coalescing, but this is only valid in aqueous systems with negligible or low ionic strength. An emulsion droplet or a liposome with charges is destabilized when electrolyte (e.g., NaCl) is added. The destabilization by electrolyte is predicted for many systems by the DLVO theory, which takes into account the competition between van der Waals attraction and electrostatic double-layer repulsion (Israelachvili, 1985).

1. Phospholipid vesicles and liposomes

Liposomes constitute an important field of research, both from an academic and an industrial perspective. Liposome technology offers great potential for various applications in areas such as drug delivery and targeting, diagnostics, skin therapies and cosmetics. Liposomes are also common model systems for cell membranes. Various aspects of liposomes have been investigated and explored in detail since their discovery by Bangham and co-workers in the early 1960s (Bangham *et al.*, 1965). Many publications and treatises exist in the area, which has been reviewed by, for example, Ostro (1987), Betageri and co-workers (1993), Lasic (1993, 1998), Lian & Ho (2001), Crommelin and co-workers (2002) and Torchillin (2005). A recent and very comprehensive review in three parts covers all aspects of liposome technology, including liposome preparation and analysis (Gregoriadis, 2006).

The classification of liposomes and various procedures for the preparation and characterization of liposomes are particularly well described in the book by Betageri and co-workers (1993), who also address the aspects of industrial applications of liposomes, but many other useful sources may be found in the literature.

Liposomes are closed vesicles of polar lipid bilayers or lamellae with enclosed aqueous layers and cores, which are able to encapsulate hydrophobic, amphiphilic and hydrophilic substances into their structure. Normally, phospholipids are used as the major polar lipid component, particularly PC, but other building blocks such as digalactosyldiacylglycerol are also known (Blom *et al.*, 1996; Carlsson *et al.*, 1996). These more or less spherical lipid particles are formed on dilution with water of a lamellar liquid crystalline phase ($T>T_m$). Liposomes are dispersions and thus thermodynamically unstable and will eventually separate into two macroscopic phases, principally the lamellar gel and/or liquid crystalline phase plus excess aqueous solution. Under certain conditions, however, it is more favourable for closed bilayers (vesicles) rather than planar bilayers to form. By this the energetically unfavourable edges are eliminated at finite, rather than infinite, aggregation number, which is entropically favoured (Israelachvili, 1985).

The easiest way to prepare multilamellar liposomes (MLV) is to hydrate a dry phospholipid mixture, preferably a thin film, take advantage of the spontaneous process of myelin formation and then subject the crude mixture to vigorous agitation by vortexing. If more than one lipid other than the principal phospholipid (PC) is used, molecular mixing is accomplished by melting (if possible) or by using a volatile organic solvent, which is then removed in a rotary evaporator. This process also promotes the formation of a thin film. The conditions of hydration and subsequent agitation are critical and must be controlled carefully to get reproducible liposomes, particularly with respect to particle size distribution and encapsulation. The multilamellar liposomes may then be further processed by ultrasonication or high-pressure homogenization in order to prepare small unilamellar liposomes (SUV) which are less prone to sediment and fuse and thus possess longer shelf life (but also possess lower encapsulation efficiency of incorporated water-soluble substances). For more details on the various preparation techniques, the reader is referred to, for example, Betageri and co-workers (1993), Lasic (1993) and Gregoriadis (2006).

The physical properties of phospholipids will influence the characteristics of the resulting liposomes. By a careful selection of the phospholipids and by including appropriate additives, it is possible to design liposomal systems which offer long shelf lives, low leakage of incorporated substances and do not undergo aggregation and sedimentation, but also liposomes with long circulating times when injected into the human bloodstream (cf. surface-modified liposomes such as polyethylene glycol-coated liposomes; Torchilin, 2005). The general rule of thumb is to select phospholipids which maintain the curved bilayer structure, through lipid shape (truncated cone formed by PC, PS, PG, sphingomyelin; see Section D), lipid charge (increases swelling and thus entrapping efficiency) and phase transition (storing at $T>T_m$ prevents fusion but increases permeability of entrapped water-soluble substances). The selection is sometimes associated with a number of compromises: well-defined synthetic phospholipids are preferred from a regulatory point of view, whereas natural phospholipids, which generally are mixtures, are preferred for reasons of both economics and performance. Saturated synthetic phospholipids and fully hydrogenated phospholipids are preferred from a chemical stability perspective, whereas unsaturated phospholipids are preferred from a colloidal stability position. Since liposomes are prepared at $T>T_m$, the use of phospholipids with phase transitions below room temperature are preferred over saturated phospholipids, due to ease of manufacture and less heat stress on the lipid material. However, $T>T_m$ is also associated with high fluidity and thus high permeability across the bilayer, but this may be overcome by the addition of high amounts of cholesterol which results in closer packing of the lipids in the membrane.

Addition of a negatively charged phospholipid such as PG may stabilize the liposomes electrostatically. This is only valid when the phospholipid mixture is dispersed in pure water or aqueous solutions with negligible or low ionic strength, since high concentrations of oppositely charged counter-ions, particularly divalent ones, diminish or completely screen the repulsive forces between the particles. It is therefore common to prepare liposomes and other lipid particles in salt-free media. Normally, a few per cent of a charged lipid is enough to cause a stabilization. Higher amounts cause 'infinite' swelling of the bilayers which is not suitable, at least not for the preparation of multilamellar vesicles; that is, vesicles with more than one closed bilayer. In case an aqueous salt solution is required, a water-soluble, non-ionic polymer such as polyethylene glycol may be added as a steric stabilizer which prevents the liposomes from aggregating.

The aqueous compartments (the enclosed aqueous layers and the cores) of a liposomal system may act as a reservoir for water-soluble molecules. The surface of the liposomal bilayer may operate as a matrix for amphiphilic molecules. The core of the liposomal bilayer may act as a solvent for non-polar molecules. Liposomes are used as delivery systems for drugs and other active substances.

A small average particle size is advantageous for the colloidal stability. Particle sizing is thus one the most important quality tests to be made and techniques based on diffraction and/or scattering of laser light are commonly used for particle size measurements. During development of liposomal preparations, it is important to monitor any changes of average particle size and size distribution, particularly at accelerated conditions (storage at elevated temperatures, repeated freezing and thawing, intense shaking).

Normally, the vesicles and liposomes have a regular spherical shape, elegantly viewed in an electron microscope after freeze-fracture of the specimen. Recently, a simple technique for the determination of the shape of liposomes has been described (Sun *et al.*, 2003). The technique is based on minicapillary viscosity measurements utilizing a very small volume of a highly diluted sample. As an example, the shape parameter *a/b* of a multilamellar liposome composed of DPPC/DPPG at a molar ratio of 9:1 has been calculated to be 1.14; i.e., such a liposome is almost perfectly spherical.

2. *Reversed vesicles*

A reversed type of phospholipid vesicles also exists, in which the polar head groups are oriented inside the bilayers and where the inside and outside of the vesicles are non-aqueous phases. Reversed vesicles are essentially the dispersion of lamellar liquid crystals in non-polar solvent. Such vesicles have been described by Kunieda and co-workers (1992). For example, a combination of PC from soybean and lyso-PC in a solvent such as squalane and the addition of a small amount of water followed by ultrasonication is a useful method of forming reversed vesicles. However, these vesicles are not particularly stable and have to be stabilized in one way or another; so far they are of limited practical use.

3. *Emulsions*

Emulsions of the oil-in-water type for pharmaceutical applications, such as clinical nutrition and for the administration of lipophilic drugs, are generally based on natural lipids. The oil is typically a vegetable oil such as soybean oil, and the emulsifier is typically a phospholipid material, such as phospholipids from egg yolk (egg lecithin) or phospholipids from soybean (soy lecithin). These emulsifiers consist of mixtures of phospholipid classes, such as phosphatidylcholine and phosphatidylethanolamine, which are zwitterionic, and phosphatidylinositol, which is ionic. The physicochemical aspects of emulsifier functionality, particularly of phospholipid emulsifiers, have been addressed by Bergenståhl (1997).

Phospholipids or phospholipid mixtures with a slightly positive curvature are preferable for the emulsification and stabilization of oil-in-water emulsions (see Section D). This may be accomplished by using natural mixtures containing a small amount of a charged component, such as the charged phospholipids PA, PG, and PI, as in soy lecithin. The amount of emulsifier added should be enough to at least cover the surface of the oil droplets. When the surfaces of the oil droplets are completely covered by phospholipids (i.e., when a monolayer has been formed), one or more bilayers are formed upon the addition of further phospholipid (Larsson, 1994). These bilayers may act as additional stabilizers of the oil droplets, but may also form separate vesicles and liposomes in the system.

If an aqueous salt solution is required, as with liposomes, a water-soluble, non-ionic polymer such as polyethylene glycol or a polysaccharide derivative may be added as a steric stabilizer, which prevents the oil droplets from aggregating (Carlsson *et al.*, 1989).

Examples of emulsions of the water-in-oil type are margarines and spreads. In these 'reversed' emulsions, the water droplets are stabilized by phospholipid mixtures, preferably resulting in a slightly negative curvature (i.e., mixtures rich in PE).

H. Microemulsions

Microemulsions are homogenous mixtures of oil, water and amphiphile. Unlike the (macro)emulsions described above, microemulsions are transparent and thermodynamically stable, and thus the term microemulsion is somewhat misleading (Holmberg *et al.*, 2003). On the microscopic level, a microemulsion consists of individual domains of oil and water separated by a monolayer of amphiphile. A microemulsion may either be oil-continuous (w/o), water-continuous (o/w), or bicontinuous, depending on composition of oil, water and polar lipid, but also depending on temperature, ionic strength, and additive (co-surfactant and/or co-solvent) for a given composition. Usually, the oil is an aliphatic or aromatic hydrocarbon. Detailed knowledge of the phase behaviour, illustrated by phase diagrams, is thus crucial for the optimal use of microemulsions in practical applications.

Attempts have been made to use PC as an amphiphile for the preparation of microemulsions. Synthetic short-chain PC can form classical droplet-like oil-continuous microemulsions (reversed micellar solutions) in the quaternary system PC–water–organic solvent–alcohol. Some naturally occurring PC mixtures also form these oil-continuous microemulsions, which also may be transformed into flexible coils upon addition of additional water (Schurtenberger *et al.*, 1993). However, pure PC is slightly too lipophilic in water to form spontaneously the zero mean curvature lipid layers needed for formation of a balanced microemulsion; that is, a bicontinuous microemulsion formed at equal amounts of oil and water at the lowest possible PC concentration. This may be surmounted by addition of a short-chain alcohol, ethanol or propanol, as a co-solvent, thereby making the polar solvent water less polar or by addition of a more polar phospholipid as a co-surfactant (Shinoda *et al.*, 1991).

The acyl chains of a phospholipid are ideal hydrophobic groups because of their strongly hydrophobic as well as strongly lipophilic properties. In order to form a balanced microemulsion, these properties should be balanced by a hydrophilic head group with both strong hydrophilic and lipophobic properties (Shinoda *et al.*, 1996). This is to ascertain that the saturation concentration both in water (i.e., the CMC) and in oil is kept low, and to favour strong enrichment of the phospholipid at the interface between the water and oil domains in the microemulsion. Since the head group of PC is slightly too lipophilic, this may be overcome by addition of the more lipophobic PA, PG, PI, and PS as co-surfactants. This is again an example of how phospholipid mixtures perform better than individual, highly purified lipids in a particular application.

I. Biomembranes

The major components of cell membranes are proteins and lipids, particularly phospholipids. This is a biomembrane with multiple functions; it acts a physical barrier between the interior of the cell and the outside environment; it actively regulates the transport of various molecules in and out of the cell; and it provides a support for proteins, and other macromolecules. The proteins are embedded in the membrane in various configurations depending on type and size of the proteins, and the generally accepted model is the Singer and Nicolson (1972) fluid mosaic model. Among the human biomembranes, myelin (neuron) and erythrocyte (red blood cell) membranes have been extensively studied (Gennis, 1989). The lipid part of these biomembranes is made up principally of PC (Figure 5.18), PE, PS, sphingomyelin, all of which are insoluble, swellable polar lipids (i.e., of the Class II type; see Section D), gangliosides (erythrocytes only; Class III), and insoluble free cholesterol (Class I). The cholesterol content in these membranes is 25–27% by weight (see Figure 5.19). Normally, there is asymmetry in the distribution of the phospholipids in the membrane due to the slow transverse diffusion ('flip-flop'

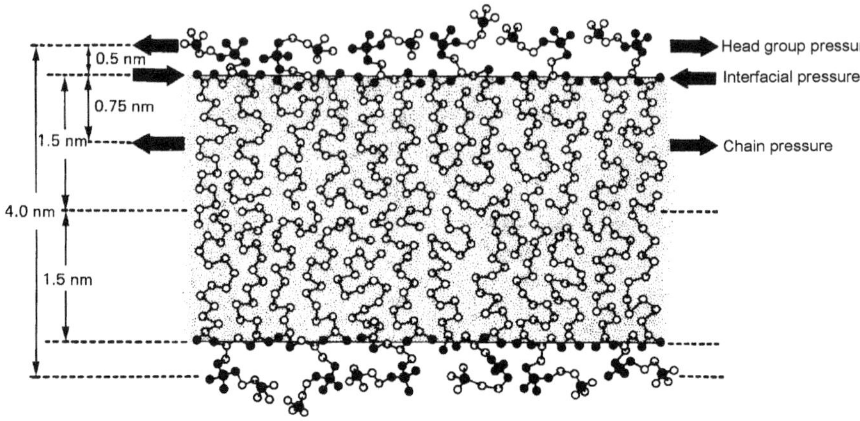

Figure 5.18. A phosphatidylcholine bilayer in its liquid crystalline state drawn to scale. The lipid bilayer is the basic structure of biological membranes. The lipids diffuse rapidly in the plane of the bilayer, covering a distance of about 1 μm in 1 s. They also cross the bilayer from one side to another ('flip-flop'), as well as exchange with lipids in the solution, but at much slower rates, of the order of hours. Reprinted from Israelachvili (1985) with kind permission of Elsevier.

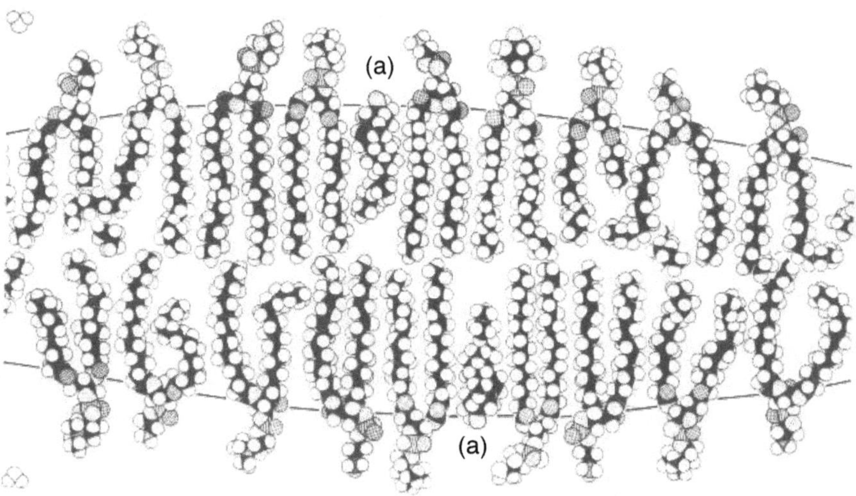

Figure 5.19. The rigid cholesterol molecule (a) can immobilize the acyl chains of adjacent phospholipids as the van der Waals attraction overcomes their kinetic movement. The molecules lengthen, narrow, and, as a result, the membrane becomes thicker. See Section F.5. Reprinted from Robertson (1983) with kind permission of Cambridge University Press.

process). The molecular geometry also plays a role, and normally more PE is found on the inside of the membrane than on the outside. The lateral diffusion (diffusion of a lipid molecule in the plane of a bilayer) is relatively high, and is directly coupled to the membrane fluidity, which in turn is kept high using phospholipids with short acyl chains and high unsaturation, as well as low cholesterol content.

Both biomembranes and model membranes with more than one lipid component may have unequal distribution of molecular components in the plane of the membrane. In addition to the embedded proteins, even the membrane lipids may distribute in a non-uniform way in the membrane. Model membranes in the liquid-ordered state (cf. Figure 5.13) may form microscopic domains rich in cholesterol and sphingomyelin, without undergoing a macroscopic phase separation. These domains are thought to be 'floating' in the membrane, without having a distinct morphological feature, and one such type of a domain has been termed a 'raft' (Epand, 2007). Such domains have attracted much interest recently, and they are extensively investigated by NMR and DSC, for example (Epand, 2007).

Another type of domain formation is the clustering of anionic lipid and protein in a fluid bilayer. Binary mixtures of zwitterionic PC and anionic PS, labelled with a small amount of a fluorescent marker, and then formulated as vesicles, in combination with a model protein (the C2 protein motif), were studied. Interpretation of fluorescence data suggested that protein binding induced anionic lipid domain until the anionic lipid became saturated with protein. The experimental conclusions were supported by Monte Carlo simulations. Thus, lipid–protein coupling is a possible mechanism for both lipid and protein clustering on a fluid bilayer (Hinderliter *et al.*, 2001).

The concept of the lipid bilayer as the structural basis of biomembranes has long been accepted, and model studies of artificial bilayers and liposomes have provided much insight into the properties of biomembranes *in vivo*. For example, the measured permeability coefficients of water and other non-electrolytes through planar and liposomal bilayers composed of PC from egg yolk are in good agreement with those for red blood cells (Gennis, 1989). The relatively high penetration of water through a model membrane in general is attributed to a low, but still existing partitioning of water in the core of the membrane. At a concentration difference of 0.1 M between the two sides of the membrane, 10–100 molecules per second per phospholipid molecule may be transported across it (Gennis, 1989). Larger non-electrolytes, such as urea and glycerol, are about 1000 times less permeable than water. Organic and inorganic electrolytes are further several orders of magnitude less permeable; whereas the proton has permeability comparable to water (Cevc, 1993). *In vivo*, the transport of solutes is also mediated by pores in the membrane, possibly induced by phase equilibria between various lipid phases in so-called tight junctions (Larsson, 1994), but also by membrane-bound proteins acting as carriers across the bilayer.

All natural lipid membranes are in the liquid crystalline state, which requires unsaturated acyl chains and/or occurrence of cholesterol, but also the presence of water. If a model bilayer is gradually dehydrated, T_m increases from temperatures well below ambient temperatures up to a point at which the existence of a gel phase is favoured. Eventually hydrated solid lipid may precipitate in excess water (see, for example, the phase diagram in Figure 5.7). Dehydration, which can cause osmotic stress and thus harm to the biomembrane, may to a certain extent be compensated for by regulation of the lipid composition by the cell. However, the osmotic pressure is strongly regulated by ions and the target for physiological saline in humans is 150 mM NaCl. In a biomembrane, small water-soluble molecules such as carbohydrates and urea also help to protect the biomembrane from undergoing an unfavourable phase transition to the gel phase. The protective effect of urea against osmotic stress in a DMPC–water system has recently been demonstrated by Costa-Balogh *et al.* (2006).

Generally, the outer surface of a cell membrane is negatively charged thanks to the presence of anionic phospholipids. Bacterial membranes in particular are heavily populated by negatively charged phospholipids, which then may interact with cationic substances. For example, cationic antimicrobial peptides exert their lytic activity mainly through electrostatic lipid–peptide interaction, which ruptures the membrane integrity (Zasloff, 2002).

Postscript

The structural lipids, notably the phospholipids, are elements of biological barriers that are amphiphilic; they have one region that associates with water (hydrophilic) and another region that rejects water (hydrophobic or lipophilic). The amphiphilic character forces these lipids to aggregate in the presence of water, which is the fundamental cause behind the formation of biological membranes: bilayers of structural lipids (phospholipids and other polar lipids) with hydrophilic regions towards the surrounding water and the lipophilic regions in the centre. The ability to self-assemble and form aggregates with hydrophilic and lipophilic properties is one major feature of phospholipids which has been dealt with in this chapter. The aim was also to provide and summarize physical data and other useful information on phospholipids, both naturally derived and chemically synthesized, in a relatively condensed format.

When I first encountered phospholipids, as an undergraduate student at the

Chemical Center, Lund University, in the early 1980s, I was quite sceptical about this amorphous material which was so different from the 'pure' chemicals I had experienced previously. Could such a sticky material, and to my naïve mind, so ill defined from a chemical perspective, ever be used in a sensible way in a chemical application? Now, more than 25 years later, I have realized that the useful properties of phospholipids indeed can be applied in a vast number of chemical applications. My own perspective on phospholipids is from the pharmaceutical side, particularly from the drug delivery point of view. Therefore, I would like to dedicate this chapter to my former supervisors in Japan, the late Professor Junzo Sunamoto, Professor Kozo Shinoda and Professor Sadao Hirota, who all have taught me much about phospholipids and their potential use in drug delivery systems.

Acknowledgements

I am indebted to Dr Dan Hallén, Dr Fredrik Carlsson and Professor Bengt Herslöf for their valuable comments on this manuscript, and to Mrs Kersti Sundbro for her kind and professional help in searching the literature.

References

Arvidson, G, Brentel, I, Kahn, A, Lindblom, G and Fontell, K (1985) Phase equilibria in four lysophosphatidylcholine/water systems. *European Journal of Biochemistry*, 152, 753–759.

Bangham, AD, Standish, MM and Watkins, JC (1965) Diffusion of univalent ions across the lamellae of swollen phospholipids. *Journal of Molecular Biology*, 13, 238–252.

Bergenståhl, B (1997) Physicochemical aspects of an emulsifier functionality. In: *Food Emulsifiers and Their Applications* (GL Hasenhuettl and RW Hartel, eds), Chapman & Hall, New York, pp. 147–172.

Betageri, GV, Jenkins, SA and Parsons, DL (1993) *Liposome Drug Delivery Systems*, Technomic Publishing Co., Lancaster.

Blom, M, Andersson, L, Carlsson, A, Herslöf, B, Zhou, L and Nilsson, Å (1996) Pharmacokinetics, tissue distribution and metabolism of intravenously administered digalactosyldiacylglycerol and monogalactosyldiacylglycerol in the rat. *Journal of Liposome Research*, 6, 737–753.

Bohlinder, K, Olsson, B, Ragnarsson, G, Singh, SK, Stjernström Öhman, G, Wadsten, T, Carlsson, A and Herslöf, B (1994) Use and characteristics of a novel lipid particle-forming matrix as a drug-carrier system. *European Journal of Pharmaceutical Sciences*, 2, 271–279.

Carlsson, A, Sato, T and Sunamoto, J (1989) Physicochemical stabilization of lipid microspheres by coating with polysaccharide derivatives. *Bulletin of the Chemical Society of Japan*, 62, 791–796.

Carlsson, A, Herslöf, B, Suurkuusk, M, Olsson, B and Singh, SK (1994) Use and characteristics of a novel lipid particle-forming matrix as a carrier system for bioactive materials. In: *Progress in Drug Delivery System III*, Biomedical Research Foundation, Tokyo.

Carlsson, A, Söderlind, E, Bergquist, M, Nilsson, R and Nylander, T (1996) Galactolipids–

a new tool in drug formulation. In: *Proceedings of the 21st World Congress of the International Society for Fat Research (ISF)*, The Hague, October 1995, PJ Barnes & Associates, Bridgwater, Vol. 2, pp. 259–262.

Cevc, G (1993) *Phospholipids Handbook*, Marcel Dekker, New York.

Chen, L, Johnson, ML and Biltonen, RL (2001) A macroscopic description of lipid–bilayer phase transitions of mixed-chain phosphatidylcholines: chain-length and chain-asymmetry dependence. *Biophysical Journal*, 80, 254–270.

Chernik, GG (1995) Phase equilibria in phospholipid–water systems. *Advances in Colloid and Interface Science*, 61, 65–129.

Costa-Balogh, FO, Wennerström, H, Wadsö, L and Sparr, E (2006) How small polar molecules protect membrane systems against osmotic stress: the urea–water–phospholipid system. *Journal of Physical Chemistry*, 110, 23845–23852.

Crommelin, DJA, Bos, GW and Storm, G (2002) Liposomes: successful carrier systems for targeted delivery of drugs. *The Drug Delivery Companies Report*, PharmaVentures, Oxford, UK.

Delogu Löwbeer, M (1994) Personal communication.

Epand, RM (2007) Detecting the presence of membrane domains using DSC. *Biophysical Chemistry*, 126, 197–200.

Fontell, K (1981) Liquid crystallinity in lipid–water systems. *Molecular Crystals and Liquid Crystals*, 63, 59–82.

Fontell, K (1992) Some aspects on the cubic phases in surfactant and surfactant-like lipid systems. *Advances in Colloid and Interface Science*, 41, 127–147.

Gennis, RB (1989) *Biomembranes: Molecular Structure and Function*, Springer-Verlag, New York.

Gregoriadis, G (2006) *Liposome Technology*, 3rd edn, CRC Press, Boca Raton.

Gunstone, FD, and Herslöf, BG (2000) *Lipid Glossary 2*, The Oily Press, Bridgwater.

Gutman, H, Arvidson, G, Fontell, K and Lindblom, G (1984) Phosphorus 31 and deuterium NMR studies of phase equilibria in the three-component system monoolein–dioleoylphosphatidylcholine–water. In: *Surfactants in Solution* (KL Mittal and B Lindman, eds), Plenum, New York, pp. 143–152.

Hinderliter, A, Almeida, PFF, Creutz, CE and Biltonen, RL (2001) Domain formation in a fluid mixed lipid bilayer modulated through binding of the C2 protein motif. *Biochemistry*, 40, 4181–4191.

Hjort Ipsen, J, Karlström, G, Mouritsen OG, Wennerström, H and Zuckermann, MJ (1987) Phase equilibria in the phosphatidylcholine–cholesterol system. *Biochimica et Biophysica Acta*, 905, 162–172.

Holmberg, K, Jönsson, B, Kronberg, B and Lindman, B (2003) *Surfactants and Polymers in Aqueous Solution*, 2nd edn, John Wiley & Sons, Chichester.

Israelachvili, JN (1985) *Intermolecular and Surface Forces – With Applications to Colloidal and Biological Systems*, Academic Press, London.

Joshi, A, Parakar, SG and Thorat, BN (2006) Modification of lecithin by physical, chemical and enzymatic methods. *European Journal of Lipid Science and Technology*, 108, 363–373.

Khan, MZI and Tucker, IG (1992) A study of the phase transition behaviours of cholesterol and saturated egg lecithin, and their interaction by differential scanning calorimetry. *Chemical and Pharmaceutical Bulletin*, 40, 3056–3061.

Koynova, R and Caffrey, M (2002) An index of lipid phase diagrams. *Chemistry and Physics of Lipids*, 115, 107–219.

Kunieda, H, Nakamura, K, Infante, MR and Solans, C (1992) Reversed vesicles from biocompatible surfactants. *Advanced Materials*, 4, 291–293.

Larsson, K (1986) Physical properties–structural and physical characteristics. In: *The Lipid Handbook* (FD Gunstone, JL Harwood and FB Padley, eds), Chapman and Hall, London, pp. 321–384.

Larsson, K (1994) *Lipids – Molecular Organization, Physical Functions and Technical Applications*, The Oily Press, Dundee.

Lasic, DD (1993) *Liposomes: From Physics to Applications*, Elsevier, Amsterdam.

Lasic, DD (1998) Novel applications of liposomes. *Trends in Biotechnology*, 16, 307–321.

Leigh, M, van Hoogevest, P and Tiemessen, H (2001) Optimising the oral bioavailability of the poorly water-soluble drug cyclosporin A using membrane lipid technology. *Drug Delivery Systems & Sciences*, 1, 73–77.

Lian, T and Ho, RJY (2001) Trends and developments in liposome drug delivery systems. *Journal of Pharmaceutical Sciences*, 90, 667–680.

Lindman, B (1987) NMR studies of structure and dynamics in surfactant systems. *Zeitschrift für Physikalische Chemie Neue Folge*, 153, 15–26.

Lindman, B and Wennerström, H (1980) Micelles. Amphiphile aggregation in aqueous solution. In: *Topics in Current Chemistry*, Volume 87, Springer-Verlag, Berlin.

Marsh, D (1990) *Handbook of Lipid Bilayers*, CRC Press, Boca Raton.

McIntosh, TJ, Simon, SA, Needham, D and Huang, CH (1992) Structure and cohesive properties of sphingomyelin/cholesterol bilayers. *Biochemistry*, 31, 2012–2020.

Ostro, MJ (1987) *Liposomes–From Biophysics to Therapeutics*, Marcel Dekker, New York.

Robertson, RN (1983) *The Lively Membranes*, Cambridge University Press, Cambridge.

Schurtenberger, P, Peng, Q, Leser, ME and Luisi, PL (1993) Structure and phase behaviour of lecithin-based microemulsions: a study of the chain length dependence. *Journal of Colloid and Interface Science*, 156, 43–51.

Seddon, JM and Cevc, G (1993) Lipid polymorphism: structure and stability of lyotropic mesophases of phospholipids. In: *Phospholipids Handbook*, Marcel Dekker, New York, pp. 403–454.

Shinoda, K, Araki, M, Sadaghiani, A, Khan, A and Lindman, B (1991) Lecithin-based microemulsions: phase behavior and microstructure. *Journal of Physical Chemistry*, 95, 989–993.

Shinoda, K, Carlsson, A and Lindman, B (1996) On the importance of hydroxyl groups in the polar head-group of nonionic surfactants and membrane lipids. *Advances in Colloid and Interface Science*, 64, 253–271.

Singer, SJ and Nicolson, GL (1972) The fluid mosaic model of the structure of cell membranes. *Science*, 175, 720–731.

Small, DM (1986) *Handbook of Lipid Research 4: The Physical Chemistry of Lipids*, Plenum Press, New York.

Söderberg, I (1990) *Structural Properties of Monoglycerides, Phospholipids and Fats in Aqueous Systems*. Doctoral Thesis, Lund University, Lund, Sweden.

Sun, Y, Li, X, Düzgüneş, N, Takaoka, Y, Ohi, S and Hirota, S (2003) The shape parameter of liposomes and DNA–lipid complexes determined by viscometry utilising small sample volumes. *Biophysical Journal*, 85, 1223–1232.

Tatulian, SA (1993) Ionization and ion binding. In: *Phospholipids Handbook*, Marcel Dekker, New York, pp. 511–552.

Torchilin, VP (2005) Recent advances with liposomes as pharmaceutical carriers. *Nature Reviews/Drug Discovery*, 4, 145–160.

Zasloff, M (2002) Antimicrobial peptides of multicellular organisms. *Nature*, 415, 389–395.

CHAPTER 6

Chemical modification

ALICE BONEKAMP

ADM Specialty Ingredients (Europe) BV, Hamburg, Germany

A. Introduction

The properties of lecithin can be modified by the chemical reaction of one or more of its components. Four such reactions are well established. These are:

- Acetylation: the reaction between acetic anhydride and the amino group of phosphatidylethanolamine (PE)
- Hydroxylation: the reaction between hydrogen peroxide and double bounds of unsaturated phospholipids
- Simultaneous acetylation and hydroxylation of lecithin
- Hydrogenation (Chapter 2, Section D.1.d).

The composition of lecithin is very complex. Fluid lecithin contains neutral lipids, such as triacylglycerols and free fatty acids, different phospholipids, mainly phosphatidylethanolamine (PE), phosphatidylcholine (PC), and phosphatidylinositol (PI) (Figure 6.1), glycolipids, carbohydrates, minerals and other minor ingredients. The reactive components can undergo different chemical reactions which modify the structure and properties of lecithin. For example, the unsaturated fatty acids present in the phospholipids can be hydroxylated. The composition of fatty acids and the degree of unsaturation vary with the source of lecithin (Table 6.1). Another reactive component is the amino group in PE. This group can be acetylated by reaction with acetic anhydride or acylated by reaction with appropriate acyl halides.

The reason for the interest in chemically modified lecithin is to raise the hydrophilic lipophilic balance (HLB) value. The products have increased hydrophilic properties compared to native lecithin, and in the case of acetylated lecithin, they also have higher colour stability through masking of the free NH_2

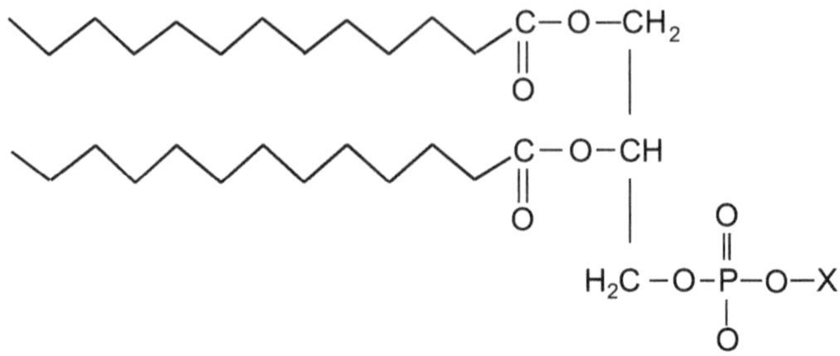

Figure 6.1. Different types of phospholipids. Phosphatidylcholine (PC), XOH = choline; phosphatidylethanolamine (PE), XOH = ethanolamine; phosphatidylinositol (PI), XOH = inositol; phosphatidic acid (PA), XOH = water.

Table 6.1. Fatty acid composition of selected plant lecithins

Acid	Acid	Soybean	Rapeseed	Sunflower	Hemp
Palmitic	16:0	18	10	14	25
Stearic	18:0	4	2	3	4
Oleic	18:1	11	39	15	11
Linoleic	18:2	58	35	61	42
Linolenic	18:3	7	8	2	13
Other		2	6	5	5

From Kuhn (1999).

group. Acetylated lecithin is resistant to heat and can be repeatedly heated and cooled with little change in colour.

Although the chemically treated lecithin is an improved oil-in-water (o/w) emulsifier, its usage in food and feed applications is not allowed in Europe, because it is not a part of the lecithin specification, E322 (E322-Lecithins, 1996). The situation is different in the US where acetylated lecithin is permitted for usage in all possible applications. However, chemically modified lecithins can be used in all industrial applications, such as leather products, paints and coatings, textiles, paper, ceramics and glass, cosmetics and soaps, inks, anti-oxidants and lubricants, and in such food applications as are allowed by food laws outside Europe. Primary food applications include,for example, use in release agents and the wetting/dispersion of beverage powders.

Hydrogenated lecithin (obtained by the reaction of unsaturated fatty acids of phospholipids and hydrogen) is more stable against oxidation and therefore is without taste and odour and has a very light colour. The main usage of hydrogenated lecithin is in the cosmetic and pharmaceutical industry (Chapter 2).

A recent review of the modification of lecithin by physical, chemical and enzymatic methods was published by Joshi and co-workers (2006). Earlier reviews of phospholipid chemistry and reactivity have been collected by Sipos and Szuhaj (1996) and by Pardun (1988). Pardun described the chemical modifications of lecithin and listed chemically modified lecithins available on the market at that time. Since then, there has been a marked concentration of companies producing and trading lecithin and today only a few companies offer chemically modified lecithin.

B. Acylated lecithin

Acylation (1) can occur at the reactive amino group of PE:

$$-NH_2 \rightarrow -NHCOR \quad \text{(acylation)} \tag{1}$$

The amino group of PE, when acylated, has a substituent on the positively charged nitrogen of the zwitterionic PE and converts the lecithin to a negatively charged species with improved solubility in water and oil-in-water emulsifying properties (Wolf, 1975). The process variables were detailed by Pardun (1988).

Doig and Diks (2003) described the process conditions. To produce a commercially usable acetylated product, the lecithin is treated with acetic anhydride (1.5–5% by wt) at 50–60°C. The resulting product is neutralized with alkali hydroxide to pH 6.5 to 8.0. The amount of acetic anhydride and the type of alkali hydroxide depend on the level of phospholipids and especially of PE in the lecithin (or gum), the degree of acetylation required and the final usage of the acetylated product. The production mixture is vacuum dried to final moisture level of maximum 1%.

$$-NH_2 \rightarrow -NHCOCH_3 \quad \text{(acetylation)} \tag{2}$$

The *N*-acetylphosphatidylethanolamine resulting from this reaction (2) is very hydrophilic and it is an effective emulsifier for o/w systems. *N*-Acetyl-PE is less sensitive to heat than normal PE and it can be repeatedly heated and cooled with minimal darkening. The relative HLB value is approximately 7 (Figure 6.2), comparable to emulsifiers like sorbitan esters.

Ghyczy (1993) reported that acylated phosphatidylethanolamines were antioxidants for unsaturated fatty acids. The chromatographic separation of soy lecithin on silica gel gave pure PE which was acylated with oleoyl chloride in Et_3N in 96% yield. Use of this product as an antioxidant for γ-linolenic acid was exemplified.

Holló (1993) discussed the possibilities for using sunflower lecithin. This product is more paste-like than natural soybean lecithin and, consequently,

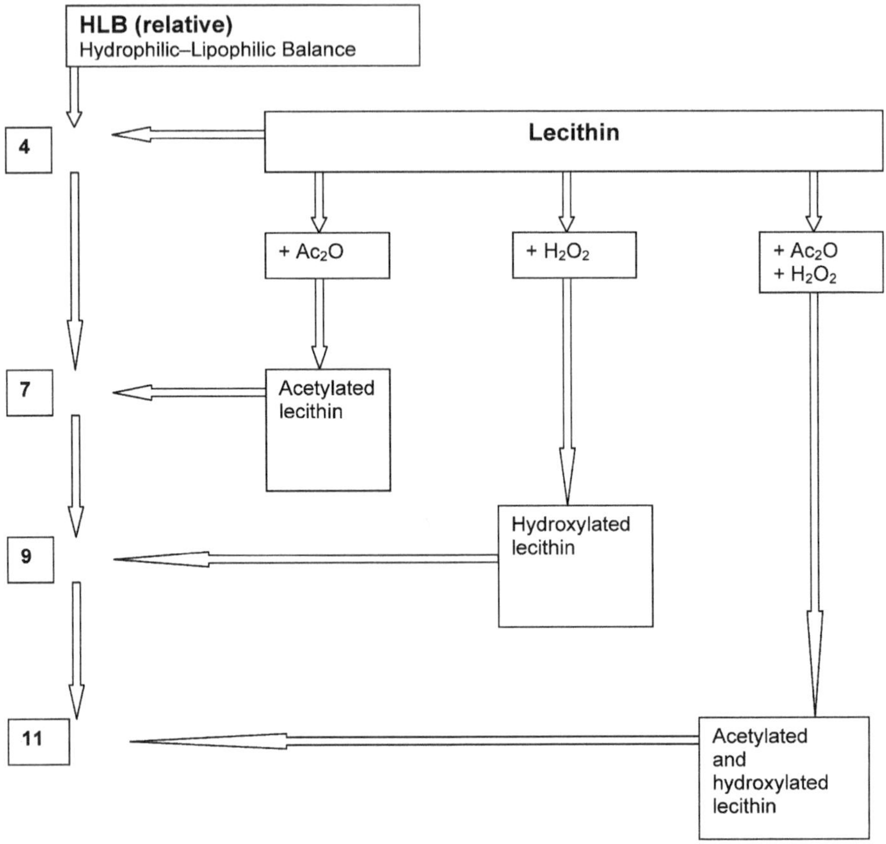

Figure 6.2. Relative HLB value and modification of lecithin. HLB: 1–6 = w/o emulsion; HLB: 8–12 = o/w emulsion.

more difficult to handle. Acylation with acid anhydride improved the fluidity and dispersibility of lecithin and enhanced its oil-in-water (o/w) emulsifying properties. Utilization of sunflower lecithin directly as an additive in food and feedstuff and after fractionation and/or modification was discussed. The acetylated products can be also used in plastic and petroleum products as reported by Wendel (2000).

Acetylated lecithin (Thermolec) is the active ingredient in many food grade release agents such as pan oils, griddle greases and aerosol coatings. In bread baking applications, lecithin acts to improve dough machinability, volume, symmetry and shelf life. In crackers, cookies, cakes and pies, lecithin improves shortening dispersion and acts as a release agent.

C. Hydroxylated lecithin

Hydroxylation is a very effective process for improving the emulsifying properties of vegetable lecithins for o/w systems. Thanks to the catalytic effect of small amounts of organic acids of low molecular weight (e.g., lactic acid), hydrogen peroxide (H_2O_2) reacts with the double bonds of unsaturated fatty acids of phospholipids and forms very hydrophilic dihydroxy fatty acid derivatives (3).

$$-CH{=}CH- \rightarrow -CH(OH)CH(OH)- \qquad (3)$$

The process has been also described by Pardun (1988), and has been published more recently by Vandana (2003) and by Joshi and co-workers (2006): a blend of crude lecithin at 50°C with lactic acid and hydrogen peroxide is stirred for about 1 h at 50°C and then dried under vacuum to a moisture content of maximum 1%. Hydroxylated lecithin is classified as GRAS (Generally Recognized As Safe) by the Food and Drug Administration under Title 21 CFR 172.814. The relative HLB value is approximately 10, depending upon the level of hydroxylation (Figure 6.2), and is comparable to emulsifiers like glycerol esters.

The usage of this lecithin is primarily in paints and in the treatment of leather (Joshi *et al.*, 2006). Lecithin has also been used in cosmetic preparations (Lau *et al.*, 1995). In the manufacture of softened cellulose fibre-based products such as paper towels, facial tissue, sanitary tissues, toilet tissues, non-woven fabrics and fluff pulp, both native and hydroxylated phospholipids were included (Evans *et al.*, 1997). The process involves treatment of cellulose fibre with chemical softening components comprising phospholipids, a non-ionic surfactant and a lubricant.

Usage of this modified lecithin in different pharmaceutical preparations is also known. Augello (2005) reported usage of hydroxylated lecithin as a part of a dispersion for spray coating of ibuprofen tablets, and as a part of prompt release coating which does not retard the release of active ingredients from the coated substrate (Augello *et al.*, 2004).

Hydroxylated lecithin (Yelkin 1018) has numerous food applications wherever superior water dispersibility is a requirement. In the baking industry, this product enhances softness, improves flavour and extends the shelf life of bread. It is used as an emulsifier in whipped and dry toppings, puddings and other foods requiring improved emulsification.

D. Acetylated and hydroxylated lecithin

The amino group of PE has a negative influence on the hydroxylation through interaction of the amino group and the performic acid (Pardun, 1988). Therefore, it is understandable that this disadvantage was overcome by combined acetylation and hydroxylation of lecithin which can produce a much more water dispersible lecithin than other products. The relative HLB value is approximately 11, depending upon the level of hydroxylation achieved (Figure 6.2), and the acetylated and hydroxylated lecithin is an excellent lecithin for many industrial applications in leather products, paper, textile, paints and pigments preparations.

The production process was described by He and Jin (1999). A procedure for modifying natural phospholipids to develop water dispersibility includes acylation, hydroxylation and neutralization. Suitable conditions for the reactions are follows: the acetylation reaction was run at 75 ± 5°C for 0.5 h, and the amount of acetic anhydride is 2–2.5% of soybean phospholipids; the hydroxylation reaction was run at 80 ± 5°C for 2 h and the amount of H_2O_2 (28%–30% aq.) was 8–10% of soybean phospholipids. Another set of process conditions was as follows: drying of soy lecithin gum at 60°C under vacuum for 10 h, acetylation with acetic anhydride at 55°C for 30 min, hydroxylation with H_2O_2 at 60–65°C for 1 h, neutralizing at 70–75°C for 30 min, and final drying (Fu, 1999). The direct treatment of soy lecithin gum was described by Fu and Lin (2001). Sagala and co-workers (2001) discussed the use of different lecithins (soy, rape, etc.) for this modification by adding peroxide while simultaneously adding carboxylic acid anhydride.

Acetylated and hydroxylated lecithin is used in the manufacture of leather, paper and textile additives, cosmetic preparations, food and medicines, paints and pigments preparations (Sagala *et al.*, 2001) and in feed or as a feed additive, which can enhance the fodder conversion, and reduce the fodder cost (Fu, 1999; Fu & Lin, 2001).

E. Industrial applications of chemically modified lecithin

Lecithins are amphiphilic substances, with an affinity to both polar and non-polar molecules. The degree to which a molecule interacts with the polar or non-polar phase depends largely upon the composition of hydrophilic and hydrophobic regions. This is expressed in the HLB value. In substances with greater HLB values, the hydrophilic region is dominant, and, therefore, these substances have more interaction with polar compounds such as water. Emulsifiers with higher HLB (8–12) tend to favour oil-in-water emulsions, while emulsifiers with lower HLB (1–6) favour water-in-oil emulsions (Colbert, 1998). The chemical modification increases the HLB value and provides products with different functionalities such as:

- Between immiscible liquid phases (i.e., oil and water), lecithin reduces surface tension and acts as an emulsifying agent
- Between solid and liquid phases, lecithin acts as a wetting and dispersing agent
- By coating the surface of solids, lecithin again reduces surface tension and the attractive forces between particles; thus, the solid is more easily dispersed and suspended in the liquid phase
- Between solid phases, lecithin can function as a lubricant and release agent.

Lecithin is used in a broad range of industrial applications including:

- Leather tanning
- Textile industry
- Paints and coatings
- Paper industry
- Plastics and ceramics
- Agrochemical emulsions.

Much of this is described in company literature, bulletins or product data sheets (Sipos, 1989).

1. Leather industry emulsifier

Leather tanning was one of the first applications for lecithin, although not understood for some time. Early man used furs and hides of slaughtered animals to make such things as clothing, shoes and tents. He observed that these skins could be made softer and suppler by rubbing the brains of the animals into the hides. In this way, the hides became an early form of untanned leather.

Skin, freed from hair, blood vessels and nerves, is agitated with lime-buffered water (pH = 12.5); it swells enormously and most of the glue-like material is extracted. On neutralization, sulfuric acid is added and the skin is treated with chromium sulfate (Bailey *et al.*, 1981). This step (tanning) involves cross-linking of protein in the hide and is a stabilization process to make the skin stiffer.

Lecithin is used in the 'fat liquoring' stage, which occurs after the tanning is complete. Fat liquors are simply oil-in-water emulsions, in which the lecithin helps form and stabilize the emulsion. To reduce the amount of fibre cohesion during the drying process and provide lubrication, the wet skin, after tanning, is treated with a fat liquor. This contains oils (sulfonated, sulfated or oxidized) and emulsifiers such as chemically modified lecithin (Sagala *et al.*, 2001; Nieuwenhuyzen, 1981). Soaps, phosphates, synthetic emulsifiers, cationic surfactants, and so on can be added. During the drying process, the emulsion droplets coalesce among the fibres, and produce soft leather. This leather can be wetted again and dried without becoming stiff. The mechanism of the softening of leather with fat liquor is not exactly clear. Different formulations for fat liquor are discussed and described by Kronick (1996).

Acetylated soy lecithin is often used in this application (e.g., Thermolec). Rapeseed lecithin modified by acetylation and hydroxylation is an excellent emulsifier.

2. *Textile industry lubricant, wetting and dispersing agent*

In the textile industry, there is a substantial demand for release agents, refatting substances, lubricants and emulsifiers, due to the very complex production processes. Although use is far from widespread, the potential for lecithin use as lubricant is tremendous. Wittcoff (1951) reported the usage of lecithin as spinning lubricant in the cotton and wool industry.

Lecithin is also able to control dust in textile manufacturing (Eichberg, 1981). In scouring, lecithins are used as wetting agents in the cleaning of natural fibres to remove fats and waxes from cotton and wool and are used as processing aids in the manufacture of synthetic fabrics (Eichberg, 1981). In dyeing, lecithins are used as wetting and dispersing agents. The presence of surfactants helps the dissolution of dye and improves the colour adsorption (Pardun, 1988).

Hydroxylated lecithin (Thermolec) has some advantages when used in textile industry. It can be completely washed off the finished goods since these molecules do not adsorb strongly on the fibre surface. Residues from unsaturated molecules, which adsorb strongly on the fibre surface and are not completely washed off the finished goods, form resinous products by autoxidation in the presence of oxygen leading to ‘yellowing’. After hydroxylation, oxidation of unsaturated centres is less and ‘yellowing’ is minimal. Therefore, the storage stability of these goods in warehouses and in the presence of ultraviolet light is improved (Kamath, 1996).

3. *Paints, coatings and inks coating aid*

Paints, coatings and inks consume a large part of the lecithin sold to the non-food/feed industries. The major advantages of lecithin in these applications are that it is functional, natural (biodegradable)(Stablec EDS) and usually cheaper than competing products. In all these applications, pigments must be ‘instantized’ (lecithin is sprayed onto the surface of pigments and incorporated into the carrier). Lecithin improves the dispersion of the pigments after surface treatment. Product benefits include improved colour development, reduced grind times, improved surface levelling, reduced foaming and improved freeze-thaw stability.

In the 1970s, Kronstein (1975, 1978) published much on the role of lecithin in paints. The paint industry has developed a wide variety of tailor-made speciality products. These are produced by blending crude or modified lecithin with selected oil carriers or other compounds to enhance lecithin's stability and usefulness in processing applications. The resulting decrease in viscosity due to added oil (Beakin LV30) or co-emulsifier (R&R551) enables lecithin to be sprayed at ambient temperature. The addition of co-emulsifiers achieves synergistic emulsification properties. Lecithin can function as an emulsifier, a wetting agent, a grinding aid, a stabilizer and a colour intensifier agent, and can compete with other surfactants on the market.

In oil-based paints, native lecithin is used as a low-cost pigment grinding aid (Nutrilec, Adlec, Yelkin). In water-based paints, modified lecithin and mixtures with other surfactants are recommended as low-cost emulsifiers, stabilizers, thickening agents and spreading aids (Sipos, 1989). A variety of formulations with modified lecithins and mixtures with other surfactants have been published (R&R551) and patented, where improved functionality in special areas are claimed. Reviews have been given by Sipos (1989) and Pardun (1988).

Mashima and Nakao (2001) described edible printing inks and their manufacture, where lecithin was used as surfactants.

4. *Paper industry softening agent*

Lecithin serves as a mixing aid in the blending of wood pulp, water, starch and other paper ingredients. In addition, lecithin's emollient characteristics give a softer, smoother texture for paper products such as facial tissues, toilet tissue and paper napkins. Modified water dispersible lecithins are generally preferred for those applications (Sagala *et al.*, 2001) and used (Thermolec 200).

5. *Plastic and ceramic industry release agent*

Release applications for lecithin in an industrial setting bear much resemblance to food and feed applications. The major difference for industrial usage is the wide range of diluents that can be chosen to produce release formulations with lecithins. Just about any carrier is allowed: hydrocarbons, vegetable oils, mineral oils, water, alcohol. Release formulations can be composed of a variety of lecithin types based upon final usage.

For mould release, product uses span the plastic/rubber, fibreglass, concrete and ceramic industries. Low temperature plastic/rubber moulding requires the use of something like native lecithin, while high temperature systems need an acetylated (heat resistant) type of lecithin. Most concrete release formulas are based on petroleum products. Lecithin is recommended not only for its solubility characteristics, but also for its anti-corrosion properties (Kronstein, 1984).

The last release application to be discussed is that for ceramics. In high performance ceramics, lecithin in dilute solution is sprayed onto the moulds (Beakin LV30). For white goods production, bleached and modified lecithin is incorporated into the ceramic powder blend. In this manner, the lecithin performs two functions – that of internal release and of improving product flow. When placed under pressure, the lecithin migrates to the surface assisting release. Additionally, the final ceramic is said to possess improved strength over those produced by other release mechanisms (Meyer, 1985).

As with mould release, there are similarities between plastic and fibreglass extrusion operations and lecithin's usage within the extrusion/pelleting operation in food/feed. Lecithin (Beakin LV30) in such operations acts as a lubricant, resulting in decreased energy consumption, decreased downtime and thus greater throughput. In many of these operations, acetylated (heat resistant) lecithin is included in the resin/product and helps in internal release.

6. *Agrochemical applications*

In agrochemical applications, lecithin utilizes all of its functionalities. In conjunction with other emulsifiers/stabilizers, such preparations generally require emulsification and dispersal of actives, reduced surface tension to allow penetration (insects or plants) of actives and enhance spraying/misting operations (Quinn, 1983). Kopp-Holtwiesche and co-workers (1995) reported the use of phospholipids for improving plant growth in which modified lecithins also were applied as a dilute aqueous preparation. Water diluted preparations of different types of chemically modified lecithin (Thermolec range) can be used for mosquito formulations.

In pesticide and fungicide applications, either native or chemically modified lecithin is used, depending upon the carrier type (aqueous or oil/solvent based). Stablec ProLec G Concentrate helps to maximize uptake/penetration of an herbicide, to reduce spray drift, and to improve the wetting properties pesticide formulation; the eco-toxicological profile is also very favourable.

F. Summary

Current concern about natural sources, sustainability and environmental issues makes the increased usage of lecithin and modified lecithin in any industrial applications more and more likely. As more vegetable oil is used as a source of bio energy, the more lecithin will be available. Lecithin as a by-product of oil processing can be easily modified chemically directly in the oil mill (gum step). This saves drying, storage and transportation costs. The chemically modified lecithin is fully compatible with synthetically emulsifiers/surfactants. This production method is shorter and easier, and perhaps the future will show that the cheaper lecithin is competitive with synthetic emulsifiers, which are more functionally defined and chemically purer.

References

Augello, M (2005) Edible propylene glycol alginate coating composition (FMC Corp, USA). USP 6932861.

Augello, M, Dell, SM, Tuason, DC, Modliszewski, JJ, Ruszkay, TA and Werner, DE (2004) Edible coating compositions containing microcrystalline cellulose and carrageenan (FMC Corp, USA). USP 6723342.

Bailey, DG, Buechler, PR and Everett, AL (1981) Leather. In: *Kirk-Othmer's Encyclopedia of Chemical Technology*, 3rd edn, Vol 14,Wiley-Interscience, NY, pp. 200–231.

Colbert, LB (1998) Lecithin tailored to your emulsification needs. *Cereal Foods World*, 43, 686–688.

Doig, SD and Diks, RMM (2003), Toolbox for modification of the lecithin headgroup. *Eur. J. Lipid Sci. Technol.*, 105, 368–376

Eichberg, J (1981) Lecithin. In: *Kirk-Othmer's Encyclopedia of Chemical Technology*, 3rd edn, Vol 14, Wiley-Interscience, NY, pp. 250–69.

Evans, RD, Morrison, S and Campbell, C (1997) Manufacture of softened cellulose fiber-based products and softening compositions (Quaker Chemical Corp, USA). PCT/US1997/002423; WO/1997/031153.

E322-Lecithins (1996) *Official J. Eur. Commun.*, L339/46, 30.12.96.

Fu, J (1999) Preparation of modified soybean lecithin for feeds (Anda Beifang Lipoid Co., Peop. Rep. China). CN 1213505 A 19990414.

Fu, J and Lin, Z (2001) Preparation of modified soybean lecithin for fodder (Anda Beifang Lipoid Co., Peop. Rep. China). CN 1295797 A 20010523.

Ghyczy, M (1993) N-Acyl-phosphatidylethanol amines as antioxidants (Rhone-Poulenc-Rorer GmbH). WO/1993/012213, DE 4141842 A1 19930624.

He, F-Y and Jin, X-J (1999), Preparation of soybean phosphatide by chemical modification. In: *Proceedings of China 8th Symposium on Industrial Surfactant Tech-Economic & Application Development*, Jingxi Huagong Bianjibu, pp. 167–169.

Holló, J, Perédi, J, Ruzics, A, Jeránek, M and Erdélyi, A (1993) Sunflower lecithin and possibilities of utilization. *Journal of the American Oil Chemists' Society*, 70, 997–1001.

Joshi, A, Paratkar, SG and Thorat, BN (2006) Modification of lecithin by physical, chemical and enzymatic methods. *Eur. J. Lipid Sci. Technol.*, 108, 363–373.

Kamath, YK (1996) Textile processing. In: *Bailey's Industrial Oil and Fat Products*, 5th edn, *Vol 1, Edible Oil and Fat Products: General Applications* (YH Hui, ed), Wiley-Interscience, New York, p. 317.

Kopp-Holtwiesche, B, Weiss, A and Schlue, R (1995) Use of phospholipids to improve plant growth (Cognis Soil Cure GmbH). WO/1996/012685, PCT/EP1995/004004 19951011.
Kronick, P (1996) In: *Bailey's Industrial Oil and Fat Products*, 5th edn, *Vol 1, Edible Oil and Fat Products: General Applications* (YH Hui, ed), Wiley-Interscience, New York, p. 309.
Kronstein, M (1975) *Modern Paint & Coatings*, 65, 1927.
Kronstein, M (1978) *Modern Paint & Coatings*, 68, 59–63.
Kronstein, M (1984) Lecithin–metaloxide compositions in corrosion preventive compounds. *Modern Paint & Coatings*, 74, 31–39.
Kuhn, M (1999) Lecithin – more than a feed additive. *Die Muhle und Mischfuttertechnik*, 136, 224–225.
Lau, JR, Geho, WB and Woods, DH (1995) Hair penetrant and carrier (SDG, Inc., USA). WO 9503773 A1 19950209.
Mashima, M and Nakao, H (2001), Edible printing inks and their manufacture (NKO Y. K., Japan). JP 20011011355 A2 200110116.
Meyer, LW (1985) Compositions containing mold release agents (Dow Chemical Co., USA). USP 4500442.
Nieuwenhuyzen van, W (1981) The industrial uses of special lecithin: a review. *Journal of the American Oil Chemists' Society*, 58, 886–888.
Pardun, HD (1988) *Die Pflanzenlecithine*, Verlag für chemische Industrie H. Ziolkowsky KG, Augsburg, pp. 265–303.
Quinn, PJ (1983) Spray method and formulation for use therein (Acacia Chemicals Ltd UK). EP 0139374
Sagala, J, Matschkal, H-P, Nagel, S and Saddington, M (2001) Emulsifier, process for its production and use (Schill & Seilacher AG, Germany). EP 1138375 A1 20011004.
Sipos, EF and Szuhaj, BF (1996) Lecithins. In: *Bailey's Industrial Oil and Fat Products*, 5th edn, *Vol 1, Edible Oil and Fat Products: General Applications* (YH Hui, ed), Wiley-Interscience, New York, p. 334.
Sipos, EF (1989) Industrial coatings applications for lecithin. In: *Lecithin: Sources, Manufacture & Uses*, (BF Szuhaj, ed) AOCS Press, Champaign, IL, p. 261.
Vandana, V, Karuna, MS Lakshmi, Prasad, PSS and Prasad, RBN (2003) Process for the preparation of hydroxylated lecithin from crude soybean lecithin. USP 6638544.
Wendel, A (2000) Lecithin: The first 150 years – part I. *Inform*, 11, 885–892.
Wittcoff, H (1951), In: *The Phosphatides*, Reinhold Publishing Corp, NY.
Wolf, WJ and Sessa, DJ (1975) *Encyclopedia of Food Science* (MS Peterson and AH Johnson, eds), AVI Publishing Co, Westport, Conn., p. 476.

CHAPTER 7

Clinical and nutritional properties of phospholipids

ASIM K. DUTTAROY

Dept of Nutrition, Faculty of Medicine, University of Oslo, Norway

A. Introduction

Phospholipids (PLs) are complex substances with chemical, biochemical and nutritional characteristics that place them in a unique category. PLs consist of fatty acids and a phosphate-containing compound attached to glycerol or the amino alcohol sphingosine, resulting in compounds with fat-soluble and water-soluble regions (Davenport, 1966) Glycerol-containing PLs include phosphatidic acid (PA), phosphatidylcholine (PC), phophatidylethanolamine (PE), phophatidylinositol (PI), phosphatidylserine (PS), sphingomyelin (SM) and sphingosine. Plasmalogens are ether phospholipids and constitute a special class of PLs characterized by the presence of an ether bond at the *sn*–1 position of glycerol rather than the usual ester bond as in diacylglycerophospholipids.

Physiologically, PLs play many essential roles in cellular structure and function (Aloj *et al.*, 1979; Barrow, 1990; Davenport, 1966; Fenske *et al.*, 1990; Petrushka *et al.*, 1959; Sala *et al.*, 1993). Their self-assembly tendency makes the PLs basic biochemical building blocks for all the known life forms. Cell membranes utilize the dual hydrophilic and hydrophobic characteristic of PLs to maintain membrane structure and function. The PLs include membrane systems on which virtually all the important metabolic reactions occur, as well as the circulating lipoproteins that transport substances around the body. However, availability of PLs may sometimes limit these essential functions. Since our bodies synthesize PLs from other nutrients and metabolites, the health significance of ingested PLs is difficult to assess. Some PLs are probably conditionally essential nutrients under some compromised conditions.

Small quantities of PLs occur naturally in a wide variety of foods, and food formulators frequently use them as emulsifiers in processed products. PLs are predominantly used as natural, non-toxic emulsifiers, wetting and dispersing agents in, for example, margarine, chocolate, bakery goods, cosmetics and pharmaceuticals. For these applications, nutritional properties of PLs do not play any role because the quantities used in food products normally range only between 0.3 to 1.0%. However, in recent decades, total dietary intake of PLs has declined, mainly due to increased refining and processing of foods. PLs have been commercially available as dietary supplements since the 1920s, and were among the first health foods (as the crude preparation, lecithin). Certain modern dietary PL supplements are clinically proven safe and are thought to be beneficial for the brain, liver, circulation and intestinal tract (Bourre *et al.*, 1993; Iwata *et al.*, 1993; Liu & Fang, 2007; Peled & Gilat, 1994). To date, health authorities recognize only choline out of these diverse compounds as a nutrient. Choline, a major component of PC and SM, functions as a methyl donor and a precursor to the neurotransmitter acetylcholine (Michel *et al.*, 2006; Miller, 1990; Zeisel, 1992). However, some consider that supplemental PLs offer some advantages in heart health and cognitive functioning. Emerging research indicates that PLs have potential for improving public health, optimizing food formulation technology and are also capable of transporting therapeutic agents into targeted tissues. In addition, they may also have beneficial effects with respect to stabilizing long-chain polyunsaturated fatty acids against oxidation in foodstuffs. This chapter will provide an account of the medical and nutrition properties of PLs.

B. Medical applications of phospholipids

1. Lung surfactants

Pulmonary surfactant is a mixture of lipids, mostly phospholipids, and proteins that permits breathing with minimal effort (Sevanian *et al.*, 1981). Surfactant phospholipids are synthesized in the type II epithelial cells of the lung. 1,2-Dipalmitoyl species of PC and phosphatidylglycerol are the main constituents of lung surfactant, which has been implicated in the maintenance of lung compliance. In order to supply sufficient amounts of palmitate for the synthesis of these dipalmitoyl species, the lung tissue actively takes up palmitate from the circulation and synthesizes palmitate PL endogenously. A lack of lung surfactant in premature infants leads to a potentially fatal condition called respiratory distress syndrome that causes shortness of breath, a major clinical problem worldwide (Friedman & Rosenberg, 1979;

Giudicelli *et al.*, 1976; Hallman, 1982; Hallman and Gluck, 1976). In addition, infection, shock, inhaling toxic chemicals and many other injuries can destroy lung surfactant or make it less effective, giving rise to the life-threatening clinical acute lung injury and acute respiratory distress syndrome (ARDS) in patients of all ages. Following lung injuries, the body releases phospholipases, which can break down the phospholipids in lung surfactant. Currently, commercial lung surfactant replacements consist of prepared mixes of phospholipids or extracts from cow or pig lung surfactant (Ballard *et al.*, 2003; Merrill & Ballard, 2003). To work well, a replacement surfactant must do several things. It must re-spread evenly after each breath. It needs to allow the lungs to transport gases such as oxygen and carbon dioxide to and from the bloodstream. It must lower surface tension in the lungs to minimize the work needed to breathe and keep the lung's air sacs alveoli from collapsing or over-expanding. The synthetic surfactants must resist being broken down by phospholipase. Future research is required in order to improve the quality of lung surfactants phospholipids.

2. *Use as liposome for drug delivery*

Liposomes have been widely investigated since 1970 as drug carriers for improving the delivery of therapeutic agents to specific sites in the body. Since liposomes are composed of phospholipids, the basic components of human cell walls, liposome encapsulation improves bioavailability of medicines, and can extend treatment effects and reduce drug dosing (Chonn & Cullis, 1995). The current pharmaceutical preparations of liposome-based therapeutic systems mainly result from our understanding of lipid–drug interactions and liposome disposition mechanisms. The success of liposomes as drug carriers is reflected in a number of liposome-based formulations, which are commercially available or are currently undergoing clinical trials. The insights gained from clinical use of liposome drug delivery systems have been integrated to design liposomes that can be targeted on tissues, cells or intracellular compartments with or without expression of target recognition molecules on liposome membranes (Harwood, 1992). Liposomes are produced by suspending a mixture of either natural or synthetic lipids/phospholipid in aqueous media (Figure 7.1) The nature of the phospholipid component(s) can vary, with neutral liposomes containing predominantly a neutral PL (e.g., PC), and cationic and anionic liposomes containing lipids that are either positively or negatively charged, respectively (Figure 7.2) (Ulrich, 2002). Liposomes of most relevance in the present context are neutral and have a spherical bilayer about 100 nm in diameter with a central aqueous core, which can be used to encapsulate any

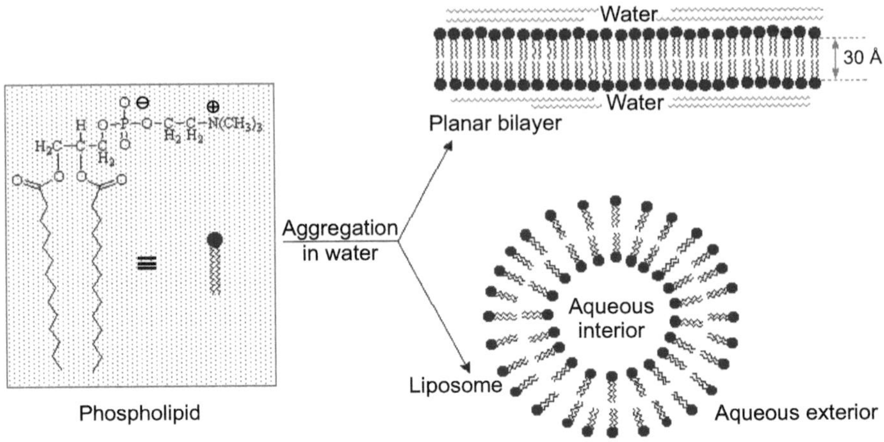

Figure 7.1. Liposomes are microscopic vesicles consisting of an aqueous core enclosed in one or more phospholipid layers. They are formed when phospholipids are vigorously mixed with water. Unlike micelles, liposomes have both aqueous interiors and exteriors.

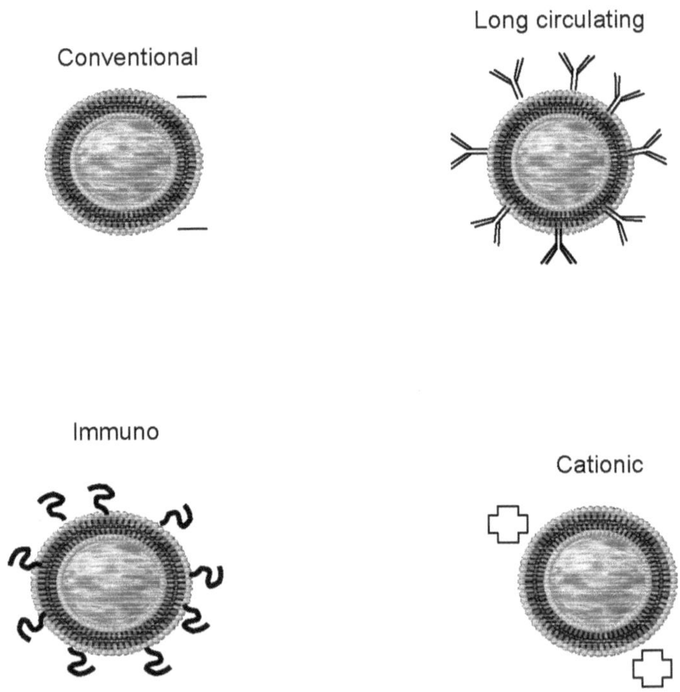

Figure 7.2. Schematic illustration of liposomes of different types.

compatible payload, such as drugs, protein or peptide antigens, or DNA (Chikh & Schutze-Redelmeier, 2002; Copland *et al.*, 2003; Ulrich, 2002).

The therapeutic advantages of liposomal drug delivery, such as the ability of long-circulating liposomes to accumulate preferentially at disease sites such as tumours (Mamot *et al.*, 2003) and sites of inflammation, are well recognized (Alpar *et al.*, 2005; Deng & Balthasar, 2005; Gregoriadis *et al.*, 2002; Liu *et al.*, 2004; Perrie *et al.*, 2003; Ritter *et al.*, 2002; Sloat & Cui, 2006; White *et al.*, 2006; Yatvin, 2002). Passive or physiological targeting of drugs to tumours is achievable using long-circulating liposomes, including pure lipid systems as well as surface-modified formulations designed to resist recognition and uptake by reticuloendothelial system cells. The pegylated liposomes are constructed with polyethylene glycol (PEG) studding the outside of the membrane. The pegylated (stealth) liposomes were used to deliver short-chain ceramide, an analogue of a naturally occurring compound that is involved in triggering cell death (Stover *et al.*, 2005). Short-chain ceramide can trigger cell death (apoptosis) in a variety of cancer cell types, but delivering this compound to tumour cells in a living animal has proven difficult. For such agents, pegylated liposome delivery offers improved chemical stability of encapsulated drug, enhanced accumulation in tumours, and prolonged drug exposure.

Further advances in liposome research have allowed liposomes to avoid detection by the body's immune system, specifically, the cells of the reticuloendothelial system. The PEG coating, which is inert in the body, allows a longer circulatory life for the drug delivery mechanism. However, research currently seeks to investigate the amount of PEG coating which hinders binding of the liposome to the delivery site. In addition to a PEG coating, most stealth liposomes also have some sort of biological species attached as a ligand to the liposome, in order to enable binding via a specific expression on the targeted drug delivery site. These targeting ligands could be monoclonal antibodies, vitamins or specific antigens. The pegylated liposomes circulate for days as stable constructs and slowly extravasate in neoangiogenic vessels in tumours, providing a degree of passive targeting to tumour tissue.

Amphotericin B is considered the treatment of choice for systemic infections caused by *Candida*, *Aspergillus* and *Fusarium*. However, it has serious side effects, particularly dose-limiting nephrotoxicity (Sabra *et al.*, 1990; Deray, 2002). Less toxic liposomal formulations of amphotericin B are used in patients with renal diseases or unresponsive to conventional amphotericin B (Sharkey *et al.*, 1996; Walsh *et al.*, 1998). Camptothecin analogues, anti-angiogenesis agents, and anti-sense oligonucleotides each represent rational candidates for delivery in highly stabilized and long-circulating liposomes. Fusogenic liposome having Sendai virus envelope glycoproteins on the surface could efficiently introduce encapsulated nucleotides and/or proteins into the cytoplasm, through its direct fusion to the plasma membrane without cytotoxicity (Kunisawa *et al.*, 2005). Recent advances in immune liposome design include rapid selection of phage antibody-derived scFv for targeting, and methods for conjugation of ligands to existing US Food and Drug Administration (FDA)-approved liposomal drugs such as pegylated liposomal doxorubicin.

Although the liposome delivery system has been tried for several drugs, only a few have been used in patients because of the slow development of necessary large-scale pharmaceutical procedures. In order to extend the degree of utilization of liposomes, future research has to focus on the production of liposomes from non-purified polar lipid sources to reduce the cost of raw materials, and to develop suitable models that can predict the rate of release of entrapped compounds. Another area of study that remains relatively unexplored is that of encapsulation of antimicrobials for long-lasting post-processing protection of ready-to-eat foods, research that is currently underway in different laboratories. The goal is to demonstrate the true potential for liposome-entrapped antimicrobials to improve the quality and safety of a wide variety of food products.

C. Roles in nutrition and health

Egg yolk, meat and internal organs are rich sources of PLs, whereas grains, fruits and vegetables are relatively low in PLs. Many of the richest dietary sources of PC and PS are also high in cholesterol and fat. Considered synonymous with PC, lecithin actually contains a mix of PLs and some triacylglycerols. For example, soy lecithin includes PA, PC, PE, PI and PS. Egg lecithin with high levels of PC qualifies as an excellent source of choline. In general, little research exists on the requirements or intake of PLs, as no known symptoms are associated with PL malnutrition, except choline deficiency. Plasma choline levels can decrease during intense physical activity of long duration. Where circulating choline concentrations have been demonstrated to decrease during physical activity, choline supplementation leads to improved performance.

Choline deficiency has been described in a variety of animal species, and manifests itself in growth, hepatic, pancreatic, lipoprotein, memory and renal abnormalities in some species (Zeisel, 2006) Choline deficiency has previously been described in patients who require long-term parenteral nutrition where it manifests in hepatic and neuropsychological abnormalities. In rodents, choline deficiency is also associated with hyperhomocysteinaemia, which is a risk factor for venous thrombosis in patients who require parenteral nutrition. Choline deficiency is a significant risk factor for development of catheter thrombosis in patients with intestinal failure (Buchman *et al.*, 2006; Henkel & Buchman, 2006).

Choline is an important nutrient throughout life. It is the precursor for the neurotransmitter acetylcholine and for PC, a structural component of very low density lipoprotein (VLDL) which is essential for normal lipid-cholesterol transport. Furthermore, choline is a source of labile methyl groups (Zeisel, 1992). For a long time, choline was considered a dispensable nutrient because it can be endogenously synthesized through sequential methylation of PE to form PC, with *S*-adenosylmethionine as the methyl donor (Ridgway & Vance, 1988; Vance & Ridgway, 1988). However, studies have shown that humans with a choline-deficient diet develop liver and kidney problems (Burt *et al.*, 1980; Buchman *et al.*, 2001)· Thus, choline is an important dietary nutrient, and an adequate intake of choline is defined as 425 mg/day for women and 550 mg/day for men.

Modern diets are depleted of phospholipids (from primarily refined oils and fats and dehulled and cleaned raw materials) These changes in diet (low fat, low cholesterol), and the BSE crisis (mad cow disease), have led to a decrease in daily PS and PC consumption. Today's foods contain approximately one third of the phospholipid levels of foods at the beginning of the last century, which could increase the athlete's need for proper nutrition and/or supplementation.

While some research indicates that choline helps with foetal brain development and memory, choline deficiency is manifested as liver dysfunction with fatty liver in studies with healthy men and people on total parenteral nutrition (da Costa *et al.*, 2004; Doggrell & Evans, 2003; Gossell-Williams & Benjamin, 2006; Innis *et al.*, 2007; Kumpf, 2006; Li *et al.*, 2006; Zeisel, 2004, 2006). Therefore, ingredients with PC, such as lecithin, and SM offer known nutritional value because they contribute choline to the diet. Lecithin is mostly a mixture of glycolipids, triacylglycerols and phospholipids (e.g., PC, PE and PI). Lecithin is commercially available in high purity as a food supplement and for medical uses.

In the dietary supplement field, health claims for lecithin and choline include improvement of exercise endurance, dementia in Alzheimer's patients, memory, thinking ability and muscle control, and reduction of liver degeneration, but with little or no scientific evidence to support them. Lecithin is regarded as a well-tolerated and non-toxic surfactant approved for human consumption by the US FDA. Lecithin is used commercially in substances requiring a natural emulsifier and/or lubricant, ranging from pharmaceuticals to protective coverings. For example, lecithin is the emulsifier that keeps cocoa and cocoa butter in a candy bar from separating. Lecithin is permitted by the EU as a food additive and designated E322. Lecithin plays an important role in the transport of fats and cholesterol from the liver to sites where they can be either used or stored. Because lecithin is required for lipoprotein synthesis, lecithin deficiency results in lipids accumulating in the liver and leads to liver damage. Lecithin deficiency also leads to increased amounts of cholesterol in the blood and atherosclerosis.

Phosphatidylserine (PtdSer) is a ubiquitous PL species that is normally located within the inner leaflet of the cell membrane. PtdSer has been implicated in a myriad of membrane-related functions. PtdSer has the potential to act as an effective antioxidant, especially in response to iron-mediated oxidation. Most of the available research that has investigated the effects of PtdSer supplementation on humans has concentrated on memory and cognitive function (Heiss *et al.*, 1994). PtdSer is commercially available as an oral supplement intended to improve cognitive function, with recommended doses usually ranging from 100–500 mg/day. The main sources that have been used to derive PtdSer for supplements are bovine-cortex and soy. However, due to the possibility of transferring infection through the consumption of prion-contaminated brain, soy sources of supplement are used in humans. Although the pharmacokinetics of PtdSer have not been fully elucidated, it is likely that oral supplementation leads to small but quantifiable increases in the PtdSer content within the cell membrane.

A small number of peer-reviewed full articles exist that investigate the effects of PtdSer supplementation in the exercising human (Kingsley *et al.*, 2006). Early research indicated that oral supplementation with PtdSer (800 mg/day) moderated exercise-induced changes to the hypothalamo-pituitary-adrenal axis in untrained participants. Subsequently, this finding was extended to suggest that PtdSer reduced the cortisol response to overtraining during weight training, while improving feelings of well-being and decreasing perceived muscle soreness (Kingsley *et al.*, 2006). However, equivocal findings suggest that the dose required to undertake this neuroendocrine action may vary between participants. Interestingly, recent findings demonstrate that short-term supplementation with PtdSer (750 mg/day) improves exercise capacity during high-intensity cycling and tends to increase performance during intermittent running. The synergistic technological and physiological properties of PtdSer make them attractive and unique candidates for the formulation of functional food.

D. Role of phospholipids in signal transmission

Over the past decade, the cell signalling roles that lipids play in the pathology and the pathophysiology of humans have been elucidated. Many types of cells methylate phospholipids using two methyltransferase enzymes that are asymmetrically distributed in membranes. As the phospholipids are successively methylated, they are translocated from the inside to the outside of the membrane (Crews *et al.*, 1981; Dainous *et al.*, 1982; Fonlupt *et al.*, 1981; Salerno & Beeler, 1973; Strittmatter *et al.*, 1981). When catecholamine neurotransmitters, lectins, immunoglobulins or chemotaxic peptides bind to the cell surface, they stimulate the methyltransferase enzymes and reduce membrane viscosity. The methylation of phospholipids is coupled to Ca^{2+} influx and the release of arachidonic acid, lysophosphatidylcholine and prostaglandins. These closely associated biochemical changes facilitate the transmission of many signals through membranes, resulting in the generation of adenosine 3',5'-monophophate in many cell types, release of histamine in mast cells and basophils, mitogenesis in lymphocytes and chemotaxis in neutrophils.

While phospholipids are major components of membranes, nature has selected lysophospholipids, a group of lipids that are more hydrophilic, to be important signalling molecules. LPA is involved in embryonic development, haemostasis, regulating stress and injury apopotosis, and cancers (Sengupta *et al.*, 2004). The importance of plasmalogens in human health is highlighted by the identification of human disorders in which plasmalogens biosynthesis is deficient. The first inborn error of metabolism in which plasmalogens were found to be deficient was the Zellweger syndrome (Brites *et al.*, 2004). For example, a hormone, neurotransmitter or growth factor secreted by one cell communicates with another by altering its cell membrane in some way, usually by activating an enzyme which breaks down phospholipid in the membrane. The breakdown products interact with an enzyme, which sets in motion a domino effect of changes in cell growth metabolism, function, and so on. Disruptions in this process may give rise to certain diseases. Some recent evidence suggests that lecithin deficiency may interfere with cell signalling and so may be a factor in the development of liver and colon cancer. Recent findings suggest a relationship between the lack of availability of lecithin to nerve cells which produce acetylcholine, and the progress of Alzheimer's disease. As a cofactor for a variety of enzymes, PtdSer is thought to be important in cell excitability and communication. PtdSer has also been shown to regulate a variety of neuroendocrine responses that include the release of acetylcholine, dopamine and noradrenaline. Additionally, PtdSer has been extensively demonstrated to influence tissue responses to inflammation.

E. Importance of *n*–3 fatty acids in membrane phospholipids

Docosahexaenoic acid, 22:6*n*–3 (DHA) is an important constituent of the membrane phospholipids of neural structures, usually occupying the *sn*-2 position (Crawford *et al.*, 1976; Decsi & Koletzko, 2005; Lauritzen *et al.*, 2001; Tam & Innis, 2006). A typical example is PE, which is especially rich in the brain and retina. Other phospholipids in which DHA is a prominent feature include PC, PI, PS, cerebrosides and sphingomyelin. There are two critical periods for the acquisition of these *n*–3 fatty acids: during foetal development, and after birth (postnatal) until the biochemical development in the brain and retina is completed. During pregnancy, both maternal stores and dietary intake of *n*–3 fatty acids are of importance in insuring that the foetus has adequate amounts of *n*–3 fatty acids at the time of birth (Crawford *et al.*, 1976; Decsi & Koletzko, 2005; Lauritzen *et al.*, 2001; Tam & Innis, 2006). DHA is preferentially transferred across the placenta into foetal blood (Duttaroy, 2000). In humans, it was shown that the administration of fish oil or sardines to pregnant women led to higher DHA concentrations in both maternal plasma and red blood cells, and in cord blood plasma and red blood cells at the time of birth (Cohen *et al.*, 2005; Decsi & Koletzko, 2005c; Helland *et al.*, 2006; Weiler *et al.*, 2005). Once membrane phospholipids have adequate concentrations of DHA, there is an avid retention of these fatty acids in the brain and the retina, even though the diet may subsequently be deficient in this acid. However, in human infants the results have been more variable and, obviously, the experimental protocols less rigorous because of ethical considerations. Even so, most studies of premature infants have shown visual impairment and abnormal electroretinograms. Recent studies in full-term infants, in which a standard infant formula was compared with human milk and with formulas enriched with DHA, provided unequivocal evidence of considerable differences in visual evoked potential (Cohen *et al.*, 2005; Decsi & Koletzko, 2005; Helland *et al.*, 2006; Weiler *et al.*, 2005). Animal model studies have shown consistent results on choline-induced memory improvements in young as well as with aged animals. The American Dietetic Association's *Health Professional's Guide to Dietary Supplements*, 2nd Edition (2003) states that claims about PS raising IQ and reducing production of cortisol, thereby inhibiting muscle breakdown, lack sufficient data. However, in 2003, the FDA approved a qualified health claim for soy-derived PS dietary supplements and the potential reduction of risk for dementia and cognitive function in the elderly, with the caveat that very limited and preliminary scientific research suggests these effects. The FDA concludes that little scientific evidence supports the claim. The FDA also specified that no level can be suggested as the effective dose.

F. Phospholipidosis

Phospholipidosis is characterized by intracellular accumulation of phospholipids with lamellar bodies, most likely from an impaired phospholipid metabolism of the lysosome. Organs affected by phospholipidosis exhibit inflammatory reactions and histopathological changes. More than 50 cationic amphiphilic drugs have been reported to induce phospholipidosis, including antidepressants, antianginal, antimalarial and cholesterol-lowering agents. Phospholipidosis has been observed as a recurrent pathological feature in toxicity studies in both animals and humans. Several cationic amphiphilic compounds are known to induce phospholipidosis; examples are amiodarone (and its major metabolite, desethylamiodarone), azithromycin, perhexiline, gentamicin, fluoxetine, chlorpromazine and tamoxifen. These drugs may induce formation of lamellar (phospholipid bilayer) inclusions in the lens and cornea of eyes (chloroquine, amiodarone), lung (amiodarone), kidney (gentamicin), or spleen lymphocytes (chlorphentermine) However, after cessation of drug administration, the phospholipidosis is reversible with drugs effluxing from cell. Di-docosahexaenoyl (22:6)-bis(monoacylglycerol)phosphate (BMP) was identified as a potential marker of drug-induced phospholipidosis. BMPs are structural isomers of phosphatidylglycerol phospholipids. BMPs are found in most cell types and tissues. They usually represent < 1% of the total phospholipid mass (Luquain *et al.*, 2001). Levels of total BMP have previously been shown to increase in serum, lymphocytes, lung and liver of amiodarone treated rats, administered with 150 mg/kg amiodarone daily for 7 days, compared to controls (Mortuza *et al.*, 2003). Cox and co-workers reported that BMP levels in liver account for 21% of the increase in phospholipids in a rat model of trospectomycin (an aminocyclitol antibiotic) induced phospholipidosis (Cox *et al.*, 1989). Despite significant advances in the understanding of overall drug-altered lipid metabolism, the relationship between impaired phospholipid metabolism and drug-induced toxicity remains enigmatic.

G. Conclusions

PLs play diverse roles in many aspects of cell biology including cell signalling and protein trafficking, cell suicide pathways, PC biosynthesis, and various issues related to choline, health, and liposomal drug delivery. PLs combine nutritional and technological properties in a single substance class. This unique and synergistic function makes them ideal candidates for use in nutrition and clinical situations. Several emerging applications of phospholipid assemblies and membrane-mimetic materials in the form of tubules, monolayers and lipid gels have been reported. Obviously, divergent views exist on the role of PL in disease prevention and treatment. Future research will provide guidance for developing PL-fortified products for preventive and therapeutic use.

References

Aloj, SM, Lee, G, Grollman, EF, Beguinot, F, Consiglio, E and Kohn, LD (1979) Role of phospholipids in the structure and function of the thyrotropin receptor. *J. Biol. Chem.*, 254, 9040–9049.

Alpar, HO, Papanicolaou, I and Bramwell, VW (2005) Strategies for DNA vaccine delivery. *Expert. Opin. Drug Deliv.*, 2, 829–842.

Ballard, PL, Merrill, JD, Godinez, RI, Godinez, MH, Truog, WE and Ballard, RA (2003) Surfactant protein profile of pulmonary surfactant in premature infants. *Am. J. Respir. Crit. Care Med.*, 168, 1123–1128.

Barrow, RE (1990) Chemical structure of phospholipids in the lungs and airways of sheep. *Respir. Physiol.*, 79, 1–8.

Bourre, JM, Dumont, O and Durand, G (1993) Brain phospholipids as dietary source of (*n*–3) polyunsaturated fatty acids for nervous tissue in the rat. *J. Neurochem.*, 60, 2018–2028.

Brites, P, Waterham, HR and Wanders, RJ (2004) Functions and biosynthesis of plasmalogens in health and disease. *Biochim. Biophys. Acta*, 1636, 219–231.

Buchman, AL, Sohel, M, Moukarzel, A, Bryant, D, Schanler, R, Awal, M, Burns, P, Dorman, K, Belfort, M, Jenden, DJ, Killip, D and Roch, M (2001) Plasma choline in normal newborns, infants, toddlers, and in very-low-birth-weight neonates requiring total parenteral nutrition. *Nutrition*, 17, 18–21.

Buchman, AL, Ament, ME, Jenden, DJ and Ahn, C (2006) Choline deficiency is associated with increased risk for venous catheter thrombosis. *JPEN J. Parenter. Enteral Nutr.*, 30, 317–320.

Burt, ME, Hanin, I and Brennan, MF (1980) Choline deficiency associated with total parenteral nutrition. *Lancet*, 2, 638–639.

Chikh, G and Schutze-Redelmeier, MP (2002) Liposomal delivery of CTL epitopes to dendritic cells. *Biosci. Rep.*, 22, 339–353.

Chonn, A and Cullis, PR (1995) Recent advances in liposomal drug-delivery systems. *Curr. Opin. Biotechnol.*, 6, 698–708.

Cohen, JT, Bellinger, DC, Connor, WE and Shaywitz, BA (2005) A quantitative analysis of prenatal intake of n–3 polyunsaturated fatty acids and cognitive development. *Am. J. Prev. Med.*, 29, 366–374.

Copland, MJ, Baird, MA, Rades, T, McKenzie, JL, Becker, B, Reck, F, Tyler, PC and Davies, NM (2003) Liposomal delivery of antigen to human dendritic cells. *Vaccine*, 21, 883–890.

Cox, JW, Ulrich, RG, Wynalda, MA, McKenna, R, Larsen, ER, Ginsberg, LC and Epps, DE (1989) Reversible, hepatic, lysosomal phospholipidosis in rat induced by subchronic daily administration of trospectomycin sulfate. *Biochem. Pharmacol.*, 38, 3535–3541.

Crawford, MA, Hassam, AG and Williams, G (1976) Essential fatty acids and fetal brain growth. *Lancet*, 1, 452–453.

Crews, FT, Calderini, G, Battistella, A and Toffano, G (1981) Age dependent changes in the methylation of rat brain phospholipids. *Brain Res.*, 229, 256–259.

da Costa, KA, Badea, M, Fischer, LM and Zeisel, SH (2004a) Elevated serum creatine phosphokinase in choline-deficient humans: mechanistic studies in C2C12 mouse myoblasts. *Am. J. Clin. Nutr.*, 80, 163–170.

Dainous, F, Freysz, L, Mozzi, R, Dreyfus, H, Louis, JC, Porcellati, G and Massarelli, R (1982) Synthesis of choline phospholipids in neuronal and glial cell cultures by the methylation pathway. *FEBS Lett.*, 146, 221–223.

Davenport, JB (1966) Structure of 'complex' phospholipids. *Nature*, 210, 198.

Decsi, T and Koletzko, B (2005) N-3 fatty acids and pregnancy outcomes. *Curr. Opin. Clin. Nutr. Metab. Care*, 8, 161–166.

Deng, R and Balthasar, JP (2005) Investigation of antibody-coated liposomes as a new treatment for immune thrombocytopenia. *Int. J. Pharm.*, 304, 51–62.

Deray, G (2002) Amphotericin B nephrotoxicity. *J. Antimicrob. Chemother.*, 49 Suppl 1, 37–41.

Doggrell, SA and Evans, S (2003) Treatment of dementia with neurotransmission modulation. *Expert. Opin. Investig. Drugs*, 12, 1633–1654.

Duttaroy, AK (2000) Transport mechanisms for long-chain polyunsaturated fatty acids in the human placenta. *Am. J. Clin. Nutr.*, 71, 315S–322S.

Fenske, DB, Chana, RS, Parmar, YI, Treleaven, WD and Cushley, RJ (1990) Structure and motion of phospholipids in human plasma lipoproteins. A 31P NMR study. *Biochemistry*, 29, 3973–3981.

Fonlupt, P, Rey, C and Pacheco, H (1981) Phosphatidylethanolamine methylation in membranes from rat cerebral cortex: effect of exogenous phospholipids and S-adenosylhomocysteine. *Biochem. Biophys. Res. Commun.*, 100, 1720–1726.

Friedman, Z and Rosenberg, A (1979) Abnormal lung surfactant related to essential fatty acid deficiency in a neonate. *Pediatrics*, 63, 855–859.

Giudicelli, Y, Pecquery, R, Lacasa, M, Boscameric, M and Nordmann, R (1976) Structure of the phosphatidylcholines of the lung surfactant at birth in normal full term infants. *Clin. Chim. Acta*, 71, 445–449.

Gossell-Williams, M and Benjamin, J (2006) Choline: are our university students eating enough? *West Indian Med. J.*, 55, 197–199.

Gregoriadis, G, Bacon, A, Caparros-Wanderley, W and McCormack, B (2002) A role for liposomes in genetic vaccination. *Vaccine*, 20 Suppl 5, B1–B9.

Hallman, M (1982) Lung surfactant phospholipids in the foetus, newborn, and in the adult; evidence of abnormality in respiratory failure. *Ann. Chir Gynaecol. Suppl.*, 196, 19–23.

Hallman, M and Gluck, L (1976) Phosphatidylglycerol in lung surfactant. III. Possible modifier of surfactant function. *J. Lipid Res.*, 17, 257–262.

Harwood, JL (1992) Understanding liposomal properties to aid their clinical usage. *Trends Biochem. Sci.*, 17, 203–204.

Heiss, WD, Kessler, J, Mielke, R, Szelies, B and Herholz, K (1994) Long-term effects of phosphatidylserine, pyritinol, and cognitive training in Alzheimer's disease. A neuropsychological, EEG, and PET investigation. *Dementia*, 5, 88–98.

Helland, IB, Saugstad, OD, Saarem, K, Van Houwelingen, AC, Nylander, G and Drevon, CA (2006) Supplementation of *n*–3 fatty acids during pregnancy and lactation reduces maternal plasma lipid levels and provides DHA to the infants. *J. Matern. Fetal Neonatal Med.*, 19, 397–406.

Henkel, AS, and Buchman, AL (2006) Nutritional support in patients with chronic liver disease. *Nat. Clin. Pract. Gastroenterol. Hepatol.*, 3, 202–209.

Innis, SM, Davidson, AG, Melynk, S and James, SJ (2007) Choline-related supplements improve abnormal plasma methionine-homocysteine metabolites and glutathione status in children with cystic fibrosis. *Am. J. Clin. Nutr.*, 85, 702–708.

Iwata, T, Kimura, Y, Tsutsumi, K, Furukawa, Y, and Kimura, S (1993) The effect of various phospholipids on plasma lipoproteins and liver lipids in hypercholesterolemic rats. *J. Nutr. Sci. Vitaminol. (Tokyo)*, 39, 63–71.

Kingsley, MI, Kilduff, L.P, McEneny, J, Dietzig, RE and Benton, D (2006) Phosphatidylserine supplementation and recovery following downhill running. *Med. Sci. Sports Exerc.*, 38, 1617–1625.

Kumpf, VJ (2006) Parenteral nutrition-associated liver disease in adult and pediatric patients. *Nutr. Clin. Pract.*, 21, 279–290.

Kunisawa, J, Masuda, T, Katayama, K, Yoshikawa, T, Tsutsumi, Y, Akashi, M, Mayumi, T and Nakagawa, S (2005) Fusogenic liposome delivers encapsulated nanoparticles for cytosolic controlled gene release. *J. Control Release*, 105, 344–353.

Lauritzen, L, Hansen, HS, Jorgensen, MH and Michaelsen, KF (2001a) The essentiality of long chain n–3 fatty acids in relation to development and function of the brain and retina. *Prog. Lipid Res.*, 40, 1–94.

Li, Z, Agellon, L.B, Allen, TM, Umeda, M, Jewell, L, Mason, A and Vance, DE (2006) The ratio of phosphatidylcholine to phosphatidylethanolamine influences membrane integrity and steatohepatitis. *Cell Metab.*, 3, 321–331.

Liu, W and Fang, DZ (2007) [A study on the changes in phospholipids content of plasma and liver in rats with hypertriglyceridemia induced by high-starch diet] Sichuan. *Da. Xue. Xue. Bao. Yi. Xue. Ban.*, 38, 129–131.

Liu, F, Shollenberger, LM and Huang, L (2004) Non-immunostimulatory nonviral vectors. *FASEB J.*, 18, 1779–1781.

Luquain, C, Laugier, C, Lagarde, M and Pageaux, JF (2001) High-performance liquid chromatography determination of bis(monoacylglycerol) phosphate and other lysophospholipids. *Anal. Biochem.*, 296, 41–48.

Mamot, C, Drummond, DC, Hong, K, Kirpotin, DB, and Park, JW (2003) Liposome-based approaches to overcome anticancer drug resistance. *Drug Resist. Updat.*, 6, 271–279.

Merrill, JD, and Ballard, RA (2003) Pulmonary surfactant for neonatal respiratory disorders. *Curr. Opin. Pediatr.*, 15, 149–154.

Michel, V, Yuan, Z, Ramsubir, S, and Bakovic, M (2006) Choline transport for phospholipid synthesis. *Exp. Biol. Med. (Maywood)*, 231, 490–504.

Miller, LG (1990) Dietary choline alteration. Implications for gamma-aminobutyric acid and other neurotransmitter receptors. *Biochem. Pharmacol.*, 40, 1179–1182.

Mortuza, GB, Neville, WA, Delaney, J, Waterfield, CJ and Camilleri, P (2003) Characterisation of a potential biomarker of phospholipidosis from amiodarone-treated rats. *Biochim. Biophys. Acta*, 1631, 136–146.

Peled, Y and Gilat, T (1994) Effect of dietary phospholipids and their constituents on bile composition in rats and hamsters. *Hepatology*, 19, 708–713.

Perrie, Y, McNeil, S and Vangala, A (2003) Liposome-mediated DNA immunisation via the subcutaneous route. *J. Drug Target*, 11, 555–563.

Petruskhka, E, Qrastel, JH and Scholefield, PG (1959) Role of phospholipids in oxidative phosphorylation and mitochondrial structure. *Can. J. Biochem. Physiol.*, 37, 989–998.

Ridgway, ND and Vance, DE (1988) Kinetic mechanism of phosphatidylethanolamine N-methyltransferase. *J. Biol. Chem.*, 263, 16864–16871.

Ritter, T, Lehmann, M and Volk, HD (2002) Improvements in gene therapy: averting the immune response to adenoviral vectors. *BioDrugs*, 16, 3–10.

Sabra, R, Takahashi, K, Branch, R.A and Badr, KF (1990) Mechanisms of amphotericin B-induced reduction of the glomerular filtration rate: a micropuncture study. *J. Pharmacol. Exp. Ther.*, 253, 34–37.

Sala, FD, Loregian, A, Lippe, G, Bertoli, E and Tanfani, F (1993) Effect of neutral and acidic phospholipids on mitochondrial ATP synthase secondary structure. *FEBS Lett.*, 336, 477–480.

Salerno, DM and Beeler, DA (1973) The biosynthesis of phospholipids and their precursors in rat liver involving de novo methylation and base-exchange pathways, in vivo. *Biochim. Biophys. Acta*, 326, 325–338.

Sengupta, S, Wang, Z, Tipps, R and Xu, Y (2004) Biology of LPA in health and disease. Semin. *Cell Dev. Biol.*, 15, 503–512.

Sevanian, A, Kaplan, SA and Barrett, CT (1981) Phospholipid synthesis in fetal lung organotypic cultures and isolated type II pneumocytes. *Biochim. Biophys. Acta*, 664, 498–512.

Sharkey, PK, Graybill, JR, Johnson, ES, Hausrath, SG, Pollard, RB, Kolokathis, A, Mildvan, D, Fan-Havard, P, Eng, RH, Patterson, TF, Pottage, JC, Jr, Simberkoff, MS, Wolf, J, Meyer, RD, Gupta, R, Lee, LW and Gordon, DS (1996) Amphotericin B lipid complex compared with amphotericin B in the treatment of cryptococcal meningitis in patients with AIDS. *Clin. Infect. Dis.*, 22, 315–321.

Sloat, BR and Cui, Z (2006) Evaluation of the immune response induced by a nasal anthrax vaccine based on the protective antigen protein in anaesthetized and non-anaesthetized mice. *J. Pharm. Pharmacol.*, 58, 439–447.

Stover, TC, Sharma, A, Robertson, GP and Kester, M (2005) Systemic delivery of liposomal short-chain ceramide limits solid tumor growth in murine models of breast adenocarcinoma. *Clin. Cancer Res.*, 11, 3465–3474.

Strittmatter, WJ, Hirata, F and Axelrod, J (1981) Regulation of the beta-adrenergic receptor by methylation of membrane phospholipids. *Adv. Cyclic. Nucleotide Res.*, 14, 83–91.

Tam, O and Innis, SM (2006) Dietary polyunsaturated fatty acids in gestation alter fetal cortical phospholipids, fatty acids and phosphatidylserine synthesis. *Dev. Neurosci.*, 28, 222–229.

Ulrich, AS (2002) Biophysical aspects of using liposomes as delivery vehicles. *Biosci. Rep.*, 22, 129–150.

Vance, DE and Ridgway, ND (1988) The methylation of phosphatidylethanolamine. *Prog. Lipid Res.*, 27, 61–79.

Walsh, TJ, Hiemenz, J.W, Seibel, N.L, Perfect, J.R, Horwith, G, Lee, L, Silber, J.L, DiNubile, MJ, Reboli, A, Bow, E, Lister, J and Anaissie, EJ (1998) Amphotericin B lipid complex for invasive fungal infections: analysis of safety and efficacy in 556 cases. *Clin. Infect. Dis.*, 26, 1383–1396.

Weiler, H, Fitzpatrick-Wong, S, Schellenberg, J, McCloy, U, Veitch, R, Kovacs, H, Kohut, J and Kin, YC (2005) Maternal and cord blood long-chain polyunsaturated fatty acids are predictive of bone mass at birth in healthy term-born infants. *Pediatr. Res.*, 58, 1254–1258.

White, K, Rades, T, Kearns, P, Toth, I and Hook, S (2006) Immunogenicity of liposomes containing lipid core peptides and the adjuvant. *Quil A. Pharm. Res.*, 23, 1473–1481.

Yatvin, MB (2002) A liposome model to treat pseudomonas in lungs of cystic fibrosis and immune compromised patients. *Cell Mol. Biol. Lett.*, 7, 304.

Zeisel, SH (1992) Choline: an important nutrient in brain development, liver function and carcinogenesis. *J. Am. Coll. Nutr.*, 11, 473–481.

Zeisel, SH (2004) Nutritional importance of choline for brain development. *J. Am. Coll. Nutr.*, 23, 621S–626S.

Zeisel, SH (2006) Choline: critical role during fetal development and dietary requirements in adults. *Annu. Rev. Nutr.*, 26, 229–250.

CHAPTER 8

Sphingolipids

ÅKE NILSSON

Lund University, Lund, Sweden

A. Introduction

Sphingolipids are minor dietary lipids which are not essential for growth or energy supply, although a source of choline and fatty acids. Due to their slow digestion by the action of mucosal rather than pancreatic enzymes and generation of the biologically active metabolites ceramide, sphingosine and sphingosine-1-phosphate (S1P), they may still have important biological effects. In the gut lumen they interact with sterols to inhibit cholesterol absorption, and they may influence triacylglycerol metabolism by unknown mechanisms. In rodents, dietary sphingolipids have antitumour effects in the gut. In the adult human diet, the most important sphingolipid sources are meat, dairy products, eggs, soy products and wheat flour. Human milk contains more sphingolipids than the milk replacement formulations. Milk gangliosides may have anti-infectious effects and sphingolipids may affect mucosal maturation in the neonate. Procedures for enrichment of milk sphingolipids in dairy products without solvent extraction are now available and products are ready for human testing. This chapter deals with the biological effects of dietary sphingolipids and their potential use in functional foods.

B. Sphingolipids in nature

1. General structure of sphingolipids

All sphingolipids share the same fundamental structure, with a ceramide unit and a polar head group. The ceramide part of the molecule consists of a sphingoid base, mostly with 18 carbon atoms, having alcohol groups at

functions (e.g., in cell signalling) of sphingolipids in plants is rapidly expanding (Sperling & Heinz, 2003).

Fungi and yeast contain CPI as a major sphingolipid. In addition, most fungi contain glucosyl and galactosylceramides (Vesper, 1999). Sphingolipid metabolism in yeast has specific features with a key role in resistance to heat and osmotic changes.

4. *Sphingoid base and fatty acid composition*

The sphingoid bases are long-chain, mainly 18 carbon atom, amino alcohols with two (d) or three (t) hydroxy groups, without or with a double bond at position 4 and/or 8 (Figure 8.2). Mammalian sphingolipids contain sphingosine which has a double bond at position 4 (d18:1^4) as the major sphingoid base but also some sphinganine (d18:0), and in some tissues phytosphinganine (t18:0), as well as smaller amounts of other sphingoid bases with 16–20 carbon atoms. In plants, the sphingoid base composition is more variable, being composed of up to eight different 18 carbon sphingoid bases (Figure 8.2). Major plant sources of sphingolipids that are of interest in a nutritional context, such as rice bran, wheat, maize and soybean glucosylceramides, contain a high proportion of 4,8-sphingadiene (d18:$2^{4,8}$, sphinga-4,8-dienine) (Sullards *et al.*, 2000). These plants also contains 8-sphingenine (d18:1^8) and 4-hydroxy-8-sphingenine (t18:1^8) whereas sphingosine (d18:1^4) is virtually absent and sphinganine (d18:0) and phytosphinganine (t18:0) are only present in minor proportions (Figure 8.2) (Sperling & Heinz, 2003). In yeast, *S. cerevisiae*, the predominant long-chain base is phytosphinganine (t18:0).

Mammalian and plant sources of sphingolipids also differ in their fatty acid composition. Animal sphingolipids contain mainly saturated long-chain fatty acids with 16–30 carbons, some of which are α-hydroxylated. Dairy sphingolipids contain primarily saturated non-hydroxylated fatty acids including odd numbered fatty acids. Milk sphingomyelin was shown to contain 2–4% 14:0, 26–36% palmitic acid (16:0), 9–12% stearic acid (18:0), 10–18% 22:0, 13–17% 23:0 and 9–14% 24:0, some 16:1, 18:1 and 24:1 (*n*–9), and small amounts of polyunsaturated fatty acids (Bitman and Wood, 1990). The fatty acids in plants are almost exclusively α-hydroxylated and vary in chain length from 14 to 26 carbon atoms. Saturated 16, 20, 22 and 24 carbon α-hydroxylated fatty acids dominate and *n*–9-monounsaturated very long-chain fatty acids ranging from 22 to 26 carbons occur in low proportions. Almost all possible combinations of the mentioned long-chain bases and fatty acids occur in plants. However, in major dietary plant sources of sphingolipids, certain molecular species dominate. For example, in wheat and soy glucosylceramide containing

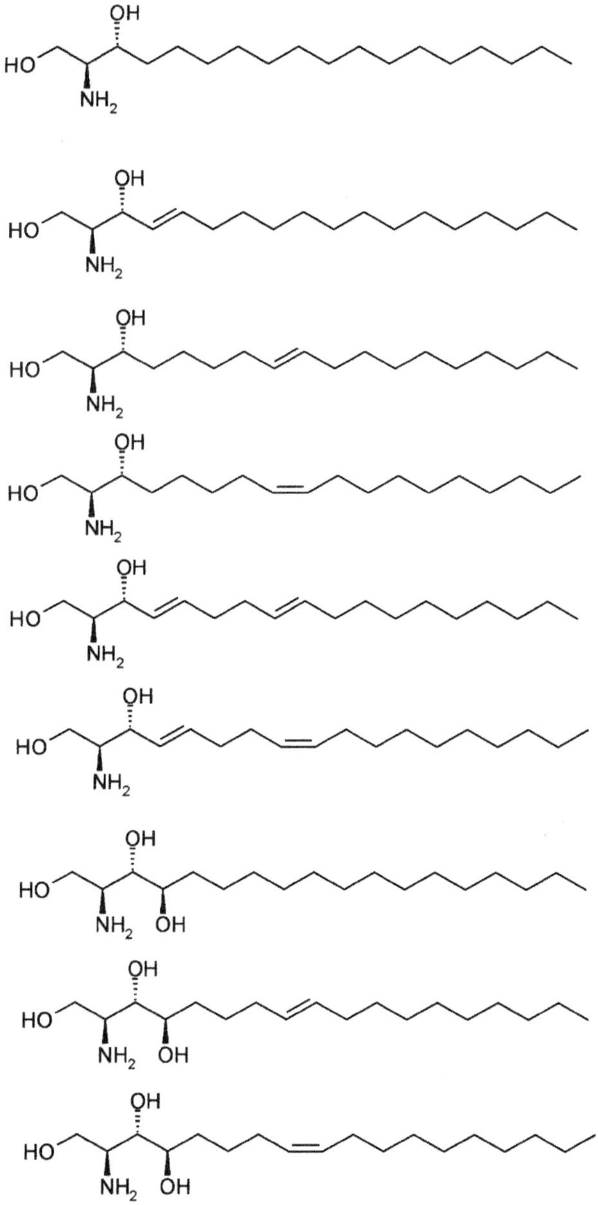

sphinganine d18:0 (dihydrosphingosine)

(*E*)-sphing-4-enine d18:1[4E] (sphingosine)

(*E*)-sphing-8-enine, d18:1[8E]

(Z)-sphing-8-enine,d18:1[8Z]

(*4E,8E*)-sphinga-4,8-dienine d18:2[4E,8E]

(*4E,8Z*)-sphinga-4,8-dienine d18:2[4E,8Z]

4-hydroxysphinganine, t18:0 (phytosphingosine)

(*E*)-4-hydroxysphing-8-enine t18:1[8E]

(*Z*)-4-hydroxysphing-8-enine t18:1[8Z]

Figure 8.2. Structure of common sphingoid bases in naturally occurring sphingolipids. Animal tissue, milk and egg sphingolipids contain mainly sphingosine, and some sphinganine and 4-hydroxysphinganine. In plant sphingolipids, the sphinga-4,8-dienines, as well as sphing-4-enine and 4-hydroxysphing-8-enines, are major sphingoid bases (Sperling & Heinz, 2003).

4,8-sphingadiene in combination with α-hydroxypalmitic acid, (h16:0) is the major species and in rice bran glucosylceramide 4,8-sphingadiene in combination with α-hydroxydecanoic acid, (h20:0) is the most prevalent species (Sullards *et al.*, 2000).

Reported analyses of plant sphingolipids are often restricted to one glycolipid class, usually glucosylceramide and/or neutral glycosylceramides as a group. Recent systematic quantitative studies indicate that the amounts of charged sphingolipids and thereby the total content of sphingolipids may sometimes have been underestimated. Results of a study by Markham and co-workers (2006), of the extraction and separation of sphingolipids from the model plant *Arabidopsis thaliana*, showed that a major polar sphingolipid is a CPI derivative. In extracts from *A. thaliana* leaves, hexose-hexuronic-CPI, monohexosylceramides and ceramides accounted for approximately 64, 34 and 2% of the total sphingolipids, respectively, suggesting an important role for the anionic sphingolipids in plant membranes. In tomatoes and soybean, a major charged sphingolipid was *N*-acetylhexosamine-hexuronic-CPI. The content of these polar sphingolipids in many plants is poorly known, and their digestion, absorption and potential biological effects have not yet been characterized.

In summary, sphingolipids are present in both animal and plant tissues and the spectrum and composition have been best characterized in animal tissues. Enrichment procedures for dairy products are already available and there is extensive knowledge about the neutral glycolipid composition of many plant components. There are thus many available sources from which sphingolipid enriched food ingredients may be prepared.

C. Sphingolipids in the diet

1. Dietary intake

The content of sphingolipids in different foods has been summarized by Vesper and co-workers (1999), Hellgren (2001, 2002) and Wehrmueller (2007). Some data from these studies, two of which also calculate the approximate total sphingolipid ingestion with an ordinary American or European diet, are summarized in Table 8.1. On the basis of the sphingolipid content of different components of the ordinary American diet, it was calculated that the daily ingestion in adults is 0.3–0.4 g, or 150–180 mmol/year (approximately 115–140 g) (Vesper *et al.*, 1999), and similar calculations based on consumption data from Switzerland indicated an average annual ingestion of about 103 g sphingolipids (Wehrmuller, 2007). The contributions of four major types of

Table 8.1. Sphingolipid content in dairy products and eggs

Product	Vesper *et al.* (1999)	Hellgren (2002)	Rombaut *et al.* (2005)	Wehrmuller (2007)
Milk 3.5%	*160*	–	–	120
Fresh milk	–	*118*	83	–
Whole milk	–	79	–	–
Low fat milk (<2%)	*92*	–	–	69 (<3.5%)
Cream (37%)	*1692*	–	–	1269
Cream (unspecified)	–	*567*	319	–
Cheese (29%)	*1326*	*192*	–	995 (>29%)
Cheddar	–	–	391	–
Quarg	–	–	104	–
Butter	*460*	*567*	520	345
Buttermilk	–	–	187	–
Eggs	*2250*	*2250*	–	1688

Figures in italics are μmol/kg; the others are mg/kg.

sources – dairy products, meat products and fish, eggs and wheat flour – are similar, and figures for vegetables and fruits are lower and more uncertain.

2. *Sphingolipids in meat, fish and dairy products*

The content of sphingolipids in fish and mammalian sources was summarized in more detail by Hellgren (2001). The contents varied between 118 (cod) and 589 (chicken leg) nmol/g. Generally, the amounts of sphingolipids were lower in fish meat than in red meat and poultry, and poultry was the richest source of this class of lipids. However, fish meat contained a higher content of neutral glycolipids than other types of meat. Thus in fish the ratio sphingomyelin/glycoslipids varied from 1 to 2.9, while in poultry this ratio varied between 5.2 and 19.2, and in red meat from 1.6 to 8.3. Poultry meat sphingomyelin also contained a higher proportion of 24:1 $\Delta 9$ and 22:1 $\Delta 9$ fatty acids than mammalian meat sphingomyelin, which is abundant in 18:0.

Ahn and Schroeder (2002) determined sphingolipid content in full fat soy flakes, soy flour, isolated soy protein, non fat dry milk, Swiss cheese and yoghurt at levels of 609 and 610 nmol/g in soy flakes and flour, 167 in cheese, 211 in soy protein, 203 in non-fat dry milk and 138 in yoghurt. Hellgren (2001) analysed the occurrence of sphingolipids in meat and fish products.

Data on the three major sphingolipids (sphingomyelin, lactosylceramide and glucosylceramide) in dairy products are summarized in Table 8.2 (Rombaut *et al.*, 2005). In agreement with the earlier finding that the proportion of sphingolipids among polar lipids in the milk fat globule membrane is high, this

of sphingolipids are now available. Combinations of extraction and enrichment procedures using natural sources with these enzymatic techniques give novel opportunities to provide formulations with optimal biological effects and less content of saturated fatty acids.

D. Digestion and absorption

1. Enzymes and metabolic pathways

Sphingolipids are not hydrolysed by pancreatic enzymes, but by mucosal enzymes that act at the brush border and in the gut lumen (Nilsson & Duan, 2006). These enzymes catalyse formation of ceramide from more complex sphingolipids and the subsequent hydrolysis of the ceramide to sphingosine and free fatty acids. *In vivo* studies in rats showed that little or no sphingolipids or ceramide are absorbed intact into blood or chyle, whereas sphingosine and sphinganine are taken up effectively by the mucosal cells (Nilsson, 1968, 1969a). Absorbed sphingoid bases are effectively converted to sphingosine-1-phosphate (S1P), most of which is rapidly metabolized to palmitaldehyde (hexadecanal) via S1P-lyase (Nilsson, 1970) (Figure 8.3). Sphingolipid digestion is slow and the whole gut thus exposed to sphingolipids and their metabolites. A small part of the absorbed sphingosine is reincorporated into ceramide and mucosal sphingomyelin and glycosphingolipids (Schmelz, 1994).

The brush border enzyme hydrolysing dietary sphingomyelin is alkaline sphingomyelinase (alk-SMase) (Nilsson, 1969b). Purification and characterization studies showed this enzyme to be related to the nucleotide/phosphodiesterase (NPP) family (Duan *et al.*, 2003a,b) and not to other sphingomyelinases. The enzyme depends on specific bile acids (taurocholate and taurochenodeoxycholate) for its action and is resistant to pancreatic proteases, making it possible for it to act under physiological conditions in the gut. Glycosylceramides are hydrolysed by the brush border lactase-phlorizin hydrolase that also hydrolyses dietary lactose (Kobayashi and Suzuki, 1981). This enzyme catalyses the formation of ceramide from glycosylceramides containing one to three hexose units. The digestion of gangliosides and other more complex glycosphingolipids is less well characterized, but a neuraminidase was found to be expressed at high levels in the middle and distal parts of the small intestine during suckling and then down-regulated to lower, adult levels (Dickson & Messer, 1978; Den Tandt *et al.*, 1987). Furthermore, bacterial glycosidases may be involved. Like other tissues, intestinal mucosa also contains acid lysososmal glycosylceramidases and sphingomyelinase to

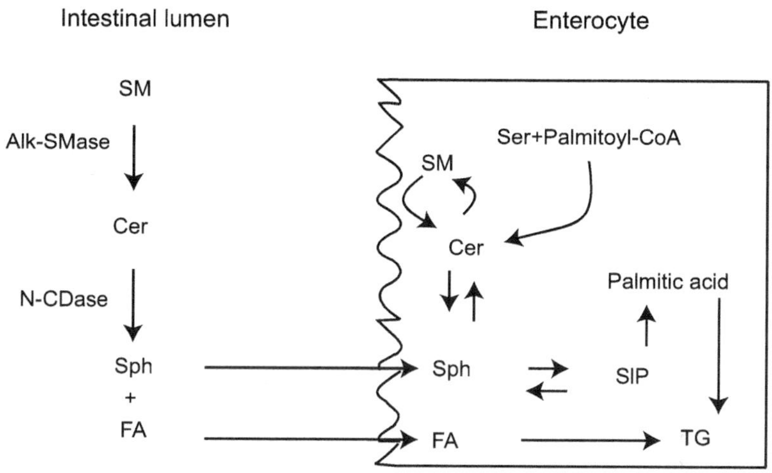

Figure 8.3. Digestion and absorption of sphingomyelin. Alk-Smase and neutral ceramidase acting at the brush border and in the gut lumen catalyse the sequential formation of sphingosine and free fatty acids which are both well absorbed. Sphingosine kinases phosphorylates sphingosine to S1P which is metabolized to palmitaldehyde and ethanolamine by S1P lyase; the palmitaldehyde is then further metabolized to palmitic acid and incorporated into chylomicron triglyceride. Some sphingosine is reacylated to ceramide

ceramide (Nilsson, 1969b). It is not clear whether these enzymes contribute to the hydrolysis of dietary sphingolipids.

Ceramide is further metabolized by a mucosal ceramidase with neutral pH optimum (Ohlsson *et al.*, 2007). This enzyme is present throughout the small intestine; i.e., it has a longitudinal distribution similar to that of alkaline sphingomyelinase (Lundgren, 2001). This enzyme is also protease resistant and active in the presence of physiological bile salt concentrations. It was purified from both rat and human small intestine and found to be identical to neutral ceramidase characterized from other tissues (Ohlsson *et al.*, 2007; Olsson *et al.*, 2004). Gene targeted animals lacking the enzyme expressed a normal phenotype but exhibited an increased faecal ceramide excretion (Kono *et al.*, 2006). Pancreatic bile salt stimulated lipase (BSSL) was shown to hydrolyse ceramide (Nyberg *et al.*, 1998). This may be of limited physiological importance *in vivo*, since gene targeting eliminating active BSSL decreased intestinal ceramidase activity only in the most upper part of the small intestine in mice, and not in the lower parts, where most of the sphingolipid digestion occurs (Kirby *et al.*, 2002).

2. *Course of sphingolipid digestion*

Whereas most of the digestion of glycerolipids occurs in the upper part of the small intestine due to the efficient action of the pancreatic lipases, sphingolipid digestion is more extended and less efficient. When varying doses of radiolabelled sphingomyelin were fed, most of the digestion occurred in the middle of the small intestine and moderate increases in the given dose extended the course of digestion and increased the exposure of the colon to undigested sphingomyelin and ceramide (Nyberg *et al.*, 1997). The digestion of glycosphingolipids appears to follow a similar course (Nilsson, 1969a). Simultaneous feeding of cholesterol or plant sterols delayed the digestion of sphingomyelin (Nyberg *et al.*, 2000). It is thus possible to increase the exposure of the whole gut to sphingolipids and their metabolites by dietary means.

Like diacylglycerol, intact long-chain ceramides are not well absorbed. Sphingosine and sphinganine, however, rapidly enter the mucosal cells. Most absorptive studies have been performed with synthetic or milk sphingolipids. Interestingly, it was shown that maize glucosylceramide, which contains mainly 4,8-sphingadiene, was hydrolysed as efficiently as glucosylceramide containing sphingosine, but the released 4,8-sphingadiene was not as well absorbed, possibly because it was more effectively expelled from the cells by the P-glycoprotein transporter (Inamine *et al.*, 2005).

In summary, the digestion of sphingolipids has unique features that make it possible to increase the exposure of the whole gut to sphingolipid metabolites by dietary feeding of sphingolipids. Human studies are few, and the digestion and absorption of ceramide phospahtidyl inositol and its derivatives have not been examined.

E. Biological effects in the gut

Studies of potentially beneficial effects of sphingolipids in the diet have focused on antitumour effects, on lipid-lowering effects, and on the ability to counteract binding of pathogenic virus and bacteria to glycosphingolipid receptors. Effects may be exerted via luminal interaction with sterols and bile salts and infectious agents, or via formation of bioactive metabolites that exert effects in the mucosa.

1. *Metabolic effects*

The ability of sphingomyelin to interact with cholesterol (Slotte, 1999) prompted a study of the effect of sphingomyelin on cholesterol absorption. Feeding sonicated mixtures of sphingomyelin and cholesterol *in vivo* and estimating the degree of cholesterol absorption by a dual isotope method indicated a lower absorption than when the radiolabelled cholesterol was fed with hydrogenated soy bean phospholipids (Nyberg *et al.*, 2000). Studying effects of egg and milk sphingomyelin on cholesterol absorption in lymphatic duct cannulated rats, Noh and Koo found a decreased cholesterol output in chyle, the effect being somewhat stronger for milk than for egg sphingomyelin (Noh & Koo, 2004). Eckhardt and co-workers studied sphingomyelin effects on cholesterol absorption in mice and in Caco 2 cells, and found an inhibitory effect related to the partition of cholesterol between micellar and monomeric form which is related to absorption kinetics for cholesterol (Eckhardt *et al.*, 2002).

Extending these findings, van Duivenvoorden and co-workers (2006) studied effects on plasma lipids and hepatic triglyceride accumulation of different types of sphingolipids in apoE Leyden mice. The regulation of plasma lipoprotein levels in this animal model exhibits several features that are similar to the human situation. Interestingly, not only sphingomyelin but also free sphingoid bases, ceramide and glycosphingolipids decreased both cholesterol and triacylglycerol levels and hepatic triacylglycerol accumulation in this animal model. This may be explained if one assumes that sphingolipids and/or sphingoid bases affect cholesterol and lipid absorption, not only by a physical interaction in the gut lumen, but also through other effects on the mucosal cell or systemic effects which have so far not been characterized. Such an influence on transporters participating in cholesterol absorption was suggested on the basis of model experiments in Caco 2 cells (Garmy *et al.*, 2005).

Results from other animal studies are somewhat contradictory. In a long-term study in normal rats, feeding sphingolipids affected the levels of plasma-cholesterol and hepatic triacylglycerol but not tissue phospholipids and sphingolipids (Kobayashi *et al.*, 1997). By feeding 1% sphingolipids to rats through two generations, a 30% decrease in plasma cholesterol was achieved and no change in plasma phospholipid classes, whereas a study over a shorter time period reported an increase in plasma SM (Imaizumi *et al.*, 1992). In a 13 week rat study with high doses of rice bran glycosphingolipids, no effect on total plasma cholesterol was seen (Moto *et al.*, 2004). Recently it was reported that feeding apoBE receptor (-/-) mice a sphingolipid-rich diet for three months increased non-HDL SM and non-HDL cholesterol (Li *et al.*, 2005). Some increase in the degree of arteriosclerosis was also observed.

In summary, animal experiments suggest that beneficial metabolic effects may be achieved by sphingolipid feeding. Further information is needed, however, particularly from human studies.

2. *Antitumour effects*

As summarized in recent reviews, research during the past 20 years has shown that the sphingolipid metabolites ceramide, sphingosine and S1P are signalling molecules which influence numerous cellular functions (Zheng *et al.*, 2006; Ogretmen & Hannun, 2004; Lahiri & Futerman, 2007). Focus has been on the ability of intracellular ceramide and sphingosine to activate pathways that favour apoptosis, cell differentiation and growth arrest, and the ability of S1P to be released from cells to act on G-protein coupled plasma membrane receptors that bind S1P at the cell surface, and activate pathways which favour cell survival and proliferation, tumorigenesis and angiogenesis. Since dietary feeding may influence the amounts of these metabolites that are formed throughout the gut, considerable interest has been devoted to antitumour effects of dietary sphingolipids in the gut, and to the enzymes that generate sphingolipid metabolites. Another important aspect is that consumption of milk seems to counteract colon cancer, and the question has been raised whether in addition to calcium and vitamin D, sphingolipids in milk may contribute to this effect (Parodi, 1997). Sphingolipids have become one of several dietary factors that may be protective in colon carcinogenesis (Lipkin *et al*, 1999).

Several studies have examined the effects of dietary sphingolipids in animal colon carcinogenesis models. The animal models used have been chemical carcinogenesis in mice induced by dimethylhydrazine (DMH) or azoxymethane (AOM). Studies have also been carried out on the *Min* mouse, which has an inactivating mutation of the FAP (APC) (familial adenomatous polyposis, adenomatous polyposis coli) gene. Like the patients with FAP, these mice develop multiple adenomas in the gut, some of which eventually develop into adenocarcinomas. The sphingolipids tested have been sphingomyelin, and glycosphingolipids from milk, maize and soybeans. The role of sphingolipids in colon cancer has recently been reviewed by Schmelz (2004) and by Duan (2005). Feeding 0.025–0.1% dairy sphingomyelin to mice treated with dimethylhydrazine decreased the number of the earliest stages in tumorigenesis (aberrant crypts) and the manifest adenocarcinomas (Schmelz *et al.*, 1997; Ahn & Schroeder, 2002). Similar studies using milk glucosylceramide and lactosylceramide (Schmelz *et al.*, 2000) or maize (Inamine *et al.*, 2005) and soy bean (Symolon *et al.*, 2004) glucosylceramide confirmed these findings. Interestingly, little difference was observed between 0.025 and 0.1% sphingolipids in the diet. In studies on *Min* mice, effects on tumour development 0.1% in the diet of a) palmitoylceramide, b) a mixture of milk sphingolipids containing the proportions of sphingomyelin, glucosylceramide, lactosylceramide and gangliosides that occur in bovine milk, or a combination of a) and b) were examined (Schmelz *et al.*, 2001). All three varieties decreased the number of tumours, the effect being strongest for the combination of a) and b). Thus the effect was sphingolipid rather than sphingomyelin specific, and even maize (Inamine *et al.*, 2005) and soy bean (Symolon *et al.*, 2004) glycosphingolipids which contain primarily 4,8-sphingadiene bases were effective.

For sphingolipids to have antitumour effects in the gut, they may have to be metabolized, since dietary ceramide was also effective (Schmelz *et al.*, 2001). The hydrolysis of dietary sphingolipids generates ceramide and sphingosine, and absorbed sphingosine will generate S1P. The longitudinal distribution and the expression of the enzymes participating in this metabolism is therefore of great interest. Alkaline SMase is expressed at highest levels in the middle and lower small intestine (Duan & Nilsson, 2000) but is found also in the colon mucosa. The enzyme activity was low in many colon tumours, in mucosa from FAP patients (Hertervig *et al.*, 1997, 1999), and in the stools of colorectal cancer patients (Di Marzio *et al.*, 2005). Purified alk-SMase was shown to inhibit proliferation of colon cancer HT29 cells (Hertervig *et al.*, 2003). This cell line has an inactivating mutation in the alk-SMase gene (Wu *et al.*, 2004). Under normal conditions, substantial amounts of alk-SMase and intestinal ceramidase reach the colon in the intestinal content. Whether these enzymes protect the colonic mucosa from tumorigenesis by generating ceramide and sphingosine in the colon is unknown.

Expression of alk-SMase can be influenced by diet; for example, it was increased by the soluble fibre psyllium and decreased by a fat-rich diet (Cheng *et al.*, 2004). Whether feeding of sphingolipids induces alk-SMase, ceramidase or lactase-phlorizine hydrolase is not known.

The rapid conversion of sphingosine to S1P via sphingosine kinase, and then to palmitaldehyde and palmitic acid via S1P-lyase, is an important feature of the metabolism of sphingosine in the small intestine (van Veldhoven and Mannaerts, 1993). S1P is both an intermediary metabolite and a messenger which acts on G-protein coupled plasma membrane receptors. It has multiple effects and tends to favour cell survival, mitogenesis and angiogenesis and may thus be an important survival messenger in tumours (Kihara *et al.*, 2007). S1P may thus counteract antitumour effects of sphingosine and ceramide and the balance between the action of sphingosine kinase, S1P lyase and S1P phosphatases that hydrolyse S1P to sphingosine and phosphate will be an important determinant of the S1P/ceramide/sphingosine ratios in tissues (Oskouian & Saba, 2007). From this perspective, it is interesting that the level of S1P-lyase is very high in the small intestine where most of the absorption of sphingosine occurs, and where tumours are rare (van Veldhoven & Mannaerts,

1993). It is also of great interest that it was recently observed that in colon tumours the S1P-lyase was shown to be down-regulated and the sphingosine kinase levels were rather high (Oskouian & Saba, 2007; Kawamori *et al.*, 2006). No experimental conditions have so far been described under which dietary sphingolipids have been shown to increase S1P levels in the intestine or colon, although this may need further study, both in normal conditions and in experimental colon carcinogenesis models.

As emphasized by Schmelz and co-workers (2001), there are several mechanisms by which dietary sphingolipids might counteract experimental colon cancer. Adding sphingosine to colon cancer cell lines inhibited growth and induced cell death and changed the intracellular distribution of β-cathenin, a protein that is a crucially important regulator of cell proliferation and which interacts with the product of the FAP gene. Sphingosine inhibits phosphorylation cycles that inhibit degradation of β-cathenin (Schmelz *et al.*, 2001). β-Cathenin is also degraded by activated caspase 3 in early apoptosis. Both sphingosine and ceramide induce caspases which are key proteases in initiation of apoptosis. The bioactive sphingolipids affect a wide spectrum of intracellular signalling pathways, however, including protein phosphatases, kinases, cAMP, concentration of intracellular calcium and the regulation of cyclooxygenases. They also interact with numerous other regulatory mechanisms such as eicosanoid metabolism, nuclear hormone receptor activation and angiogenic and inflammatory factors. In general, S1P acts as a mitogen and angiogenic growth factor, stimulating DNA synthesis, inhibiting apoptosis and inducing blood vessel formation. For example, S1P has been shown to induce COX-2 expression, thereby enhancing inflammation and tumorigenesis. Examples of a more detailed description of the relation between sphingolipid signalling, are to be found in Ogretmen and Hannun (2004), Lahiri and Futerman (2007), and El Alwani and co-workers (2006).

Phospholipids in bile prevent cytotoxicity induced by bile salts (Elferink & Groen, 1999). This effect is particularly pronounced for sphingomyelin (Moschetta *et al.*, 2000). This effect may be important in the middle and lower small intestine where some undigested sphingomyelin persists but where most other lipids, but not the bile salts, have been absorbed. Sphingomyelin was also found to prevent deoxycholate induced apoptosis and hyperproliferation in Caco 2 cells (Moschetta *et al.*, 2003). This finding raises the interesting possibility that sphingolipids may also exert an anti-carcinogenic effect in the gut by binding co-carcinogenic bile salts.

It is not known whether or not dietary sphingolipids influence inflammatory processes in the gut. Positive effects of a slow release formulation of egg phospholipids (primarily phosphatidylcholine) were reported. Studies in animal models of inflammation are therefore needed. The biological basis for expecting positive or negative effects is complex (Pettus *et al.*, 2004; El Alwani *et al.*, 2006). Interestingly, a novel inhibitor of S1P relieves experimental colitis, and SlP has been ascribed a key role in the regulation of lymphocyte trafficking (Cyster, 2007).

3. Anti-infectious effects

Some microorganisms and microbial toxins use glycosphingolipids to attach to the host cell as a way of mediating pathogenic effects. Viruses can bind to host cells via glycosphingolipids. Dietary sphingolipids may thus compete for the attachment sites and promote elimination of pathogenic organisms and toxins from the intestine and counteract bacterial translocation. Milk fat consumption was shown to be related to a reduced number of food-borne infections (Sprong *et al.*, 2001). The mechanism behind this finding, however, has not been elucidated. The bactericidal activity of sphingosine and ceramide on the pathogenic bacterial strains *E. coli* O157:H7, *Salmonella enteritidis*, *Campylobacter jejuni* and *Listeria monocytogenes* was investigated *in vitro*. It was found that sphingosine but not ceramide has a strong antibacterial effect on all strains (Sprong *et al.*, 2001). Thus in the small intestine, where digestion of sphingolipids is fastest, the continuous generation of sphingosine may be important for the control of bacterial growth.

Microorganisms, microbial toxins (Smith *et al.*, 2004; Lingwood *et al.*, 2000) and viruses that bind to cells via sphingolipids include cholera toxin (GangliosideM1), verotoxin (globosides), Shiga-like toxin 2e and *Clostridum botulinum* type B neurotoxin. Furthermore, many bacteria utilize sphingolipids to adhere to cells; for example, *E. coli* (galactosylceramide), *Hemophilus influenzae* (gangliotetraosylceramide and gangliotriaosylceramide), *H. pylori* (several), *Borrelia burgdorferii*, *Pseudomonas aeruginosa* and *Candida albicans*. Several viruses as HIV-1 gp120, Sendai virus, influenza virus and rotavirus have similar features (Rolsma *et al.*, 1998).

In conclusion, these antibacterial and antiviral mechanisms of dietary sphingolipids may be important but the *in vivo* relevance is often hard to evaluate.

4. Other health effects

Due to the important function of long-chain and very long-chain ceramides in the barrier function of stratum corneum in the skin, there is a considerable interest in ceramide formulation for treatment of skin diseases such as atopic dermatitis and psoriasis (Choi & Maibach, 2005). Sphingolipids have been used for local treatment of dermatological disorders, in order to improve barrier function. Transient positive effects have been reported and even positive effects of oral ingestion of oral glucosylceramide in atopic dermatitis (Kimata, 2006). Topical treatment of keloids with topical sphingosine was reported to be potentially beneficial, due to the induction of apoptosis in cells inducing connective tissue formation. Also, the bactericidal effect of free sphingoid bases may be of interest in topical treatments. Interestingly, it was also demonstrated that when glucosylceramide was fed orally to hairless mice, skin function was improved (Tsuji *et al.*, 2006). The mechanism behind these effects of oral glucosylceramide is unknown.

F. Sphingolipids in infant nutrition

1. Composition and metabolism of human milk sphingolipids

As a result of the high sphingolipid content in the milk fat globule membrane, the suckling neonate ingests significant amounts of sphingolipids. Like bovine milk, human milk contains sphingomyelin, lactosylceramide, glucosylceramide and gangliosides. The sphingomyelin content was found to be 120–150 mg/l (163–206 μmol/l) (Zeisel *et al.*, 1986). Per body weight, the suckling baby thus ingests more sphingolipids per day than the adult. Milk replacement formulations contain less sphingolipids than mothers' milk (Zeisel *et al.*, 1986). The question is therefore whether this matters. It has been suggested that metabolites from sphingolipids contribute favourable effects during suckling, and there is some evidence that sphingomyelin does indeed influence gut maturation in newborn rats (Motouri *et al.*, 2003). The key enzymes in sphingolipid digestion (alkaline sphingomyelinase and neutral ceramidase) are expressed as the differentiated epithelium is formed (Lillienau *et al.*, 2003) and in humans, analyses of meconium showed the presence of alkaline sphingomyelinase as well as ceramidase and sphingolipid metabolites in both term and preterm babies (Duan *et al.*, 2007). Evidence for digestion and absorption of nervonic acid (24:1) from orally given sphingomyelin, which was transferred from ingested sphingolipids to tissues in suckling mouse neonates, has been obtained (Bettger *et al.*, 2003).

2. Gangliosides in milk

Gangliosides are a complex group of sphingolipids composed of glycosylceramides containing at least one sialic acid (*N*-acetylneuraminic acid, Neu5Ac)

Table 8.4. Glycosphingolipids of common edible plant tissues

Product	Glycosylceramide (monohexosylceramide)
Cereals	
Barley	144
Corn	115
Rice (milled)	25
Wheat flour	210
Legumes	
Pea	173
Soy bean	80
Leaf vegetables	
Cabbage	100
Lettuce	51
Parsley	44
Perilla	409
Spinach	139
Fruits	
Apple	71
Banana	12
Grape	79
Grapefruit	89
Kiwi fruit	62
Pear	68
Strawberry	34
Average	101

Values are mg/kg from Sugawara and Miyazawa (1999).

joined by glycosidic linkage to other monosaccharide units. In bovine milk, these are dominated by the disialoganglioside GD3 (Neu5Ac-Neu5Ac-galactose-glucosylceramide) which, together with the monosialoganglioside GM3 (Neu5Ac-galactose-glucosylceramide), may account for more than 80% of the total ganglioside content. Using HPLC/tandem mass spectroscopy, Sorensen (2006) recently found the GD3 content to be on average 9.3 mg/l GD3 in Holstein and 17.5 mg/l in Jersey cow milk, the GM3 content being of the order 0.1–0.2 mg/l. Formula fed infants ingest only 20% of that ingested by breastfed babies (Heine *et al.*, 1993). Comparisons between human and bovine milk indicate that the content is significantly higher in human milk. The total amount of gangliosides expressed as lipid-bound sialic acid was 9.51 (1.116) mg/l (mean SD) in colostrum and 9.07 (1.15) mg/l in mature human milk. By the same method, mature bovine milk contained 3.98 (0.25) mg/l (Pan & Izumi,

2000; Sanchez-Diaz *et al.*, 1997). As a consequence, the ganglioside content in bovine milk based infant formulas is also lower than in human milk (Pan & Izumi, 2000; Sanchez-Diaz *et al.*, 1997). Furthermore, the fatty acid composition differs between human and bovine milk gangliosides (Bode *et al.*, 2004). Both contain high proportions of saturated fatty acids with 16–24 carbon atoms, but the bovine form contains much 23:0 in exchange mainly for 18:0. The pattern of gangliosides changes during lactation (Rueda *et al.*, 1995). Depending on the stage of lactation, breast milk contains up to twice as much lipid-bound sialic acid, which correlates to ganglioside content. The field of milk gangliosides has been reviewed (Rueda *et al.*, 1998a; Pan & Izumi, 2000). Human milk also contains GM1 (II^3adiNeuAc-nLcOsa$_4$Cer) and a number of highly polar gangliosides.

Gangliosides in milk have effects that may be potentially important during suckling. They may protect breast-fed infants from infections. It has been shown that milk gangliosides can inhibit enterotoxins, inhibit adhesion of *E. coli* and suppress the growth of *E. coli* and potentially pathogenic microorganisms in the intestine of premature babies and stimulate growth of bifidobacteriae. Milk gangliosides also inhibited cholera toxin adherence to Caco 2 cells (Idota & Kawakami, 1995). Enriching infant formula with ganglioside changed the microflora. *E. coli* in the faeces of preterm infants decreased and bifidobacteriaa increased, and the growth of bifidobacteria is promoted by sialic acid (Rueda *et al.*, 1998b). Gangliosides may be essential for neonatal growth of the nervous system (Wang & Brand-Miller, 2003) and may enhance uptake of lipids in the weanling rat (Birecki *et al.*, 2006).

The *in vivo* relevance of the milk ganglioside effects are difficult to evaluate but attention should be given to the ganglioside content of the milk formulas.

G. Can dietary sphingolipids have adverse effects?

The potential use of sphingolipids in functional foods postulates that no adverse effects are initiated by the enrichment of a food with sphingolipids. So far, no effects have been reported which speak strongly against this use. The observation that addition of sphingolipids to the diet increased plasma cholesterol and sphingomyelin in apo BE receptor knockout mice (Li *et al.*, 2005) should not prevent human studies, but reminds us that sensitive animal models and human individuals may react differently, and also that sphingolipids are a source of saturated fatty acids, including palmitic acid formed from the sphingoid bases. Interestingly, a 13 week broad toxicity test of rice bran glucosylceramide in rats using doses up to 1000 mg/kg body weight per day showed no toxic effects and no influence on total plasma cholesterol (Moto *et al.*, 2004). Theoretically there may be circumstances under which sphingolipid feeding increases S1P formation in the gut which might enhance inflammation and tumorigenesis, and this aspect should be further scrutinized.

H. Future perspectives

There is current evidence that dietary sphingolipids may have a number of beneficial effects, but there is still a need for additional evidence, in particular from human studies. As formulations that are enriched in sphingolipids can now be produced in a cost-effective way, knowledge is expected to expand both with regard to effects and safety aspects.

Acknowledgement

Thanks to Rui-Dong Duan and Lena Ohlsson for valuable help during the preparation of this article. The work in the authors research group eas supported by The Swedish Research Council Medicine (ÅN), the Swedish Research Council Formas (ÅN and LO) and The Swedish Cancer Society (RDD).

References

Ahn, EH (2002) Bioactive sphingolipids are constituents of soy and dairy products. *J. Food Science*, 67, 522–524.

Ahn, EH and Schroeder JJ (2002) Sphingoid bases and ceramide induce apoptosis in HT29 and HCT 116 human colon cancer cells. *Exp. Biol. Med. (Maywood)*, 227, 345–353.

Bettger WJ, DiMichelle-Ranalli, E, Dillingham, B and Blackadar, CB (2003) Nervonic acid is transferred from the maternal diet to milk and tissues of suckling rat pups. *J. Nutr. Biochem.*, 14, 160–165.

Birecki, CJ, Drozdowski, LA, Suh, M, Park, EJ, Clandinin, MT and Thomson, AB (2006) Dietary gangliosides enhance lipid uptake in weanling rats. *J. Pediatr. Gastroenterol. Nutr.*, 42, 59–65.

Bitman, J and Wood, DL (1990) Changes in milk fat during lactation. *J. Dairy Sci.*, 73, 1208–1216.

Bode, L, Beermann, C, Mank, M, Kohn, G and Boehm, G (2004) Human milk gangliosides differ in their fatty acid composition. *J. Nutr.*, 134, 3016–3020.

Cheng, Y, Ohlsson, L and Duan, RD (2004) Psyllium and fat in diets differentially affect the activities and expressions of colonic sphingomyelinases and caspase in mice. *Br. J. Nutr.*, 91, 715–23.

Choi, MJ and Maibach, HI (2005) Role of ceramides in barrier function of healthy and diseased skin. *Am. J. Clin. Dermatol.*, 6, 215–223.

Cyster, JG (2007) Chemokines, sphingosine-1-phosphate, and cell migration in secondary lymphoid organs. *Novartis Found. Symp.*, 281, 54–61; discussion 61–4, 208–9.

Den Tandt, WR, Adriaenssens, K and Scharpe, S (1987) Characteristics of human intestinal acid sialidase. *Enzyme*, 37, 155–158.

Dickson, JJ and Messer, M (1978) Intestinal neuraminidase activity of suckling rats and other mammals. Relationship to the sialic acid content of milk. *Biochem. J.*, 170, 407–413.

Di Marzio, L, Di Leo, A, Cinque, B, Fanini, D, Agnifili, A, Berloco, P, Linsalata, M, Lorusso, D, Barone, M, De Simone, C and Cifone, MG (2005) Detection of alkaline sphingomyelinase activity in human stool: proposed role as a new diagnostic and prognostic marker of colorectal cancer. *Cancer Epidemiol. Biomarkers Prev.*, 14, 856–862.

Duan, RD (2005) Anticancer compounds and sphingolipid metabolism in the colon. *In Vivo*, 19, 293–300.

Duan, RD and Nilsson, A (2000) Sphingolipid hydrolyzing enzymes in the gastrointestinal tract. *Methods Enzymol.*, 311, 276–286.

Duan, RD, Bergman, T, Xu, N, Wu, J, Cheng, Y, Duan, J, Nelander, S, Palmberg, C and Nilsson, A (2003a) Identification of human intestinal alkaline sphingomyelinase as a novel ecto-enzyme related to the nucleotide phosphodiesterase family. *J. Biol. Chem.*, 278, 38528–38536.

Duan, RD, Cheng, Y, Hansen, G, Hertervig, E, Liu, JJ, Syk, I, Sjostrom, H and Nilsson, A (2003b) Purification, localization, and expression of human intestinal alkaline sphingomyelinase. *J. Lipid Res.*, 44, 1241–1250.

Duan, RD, Cheng, Y, Jonsson, BA, Ohlsson, L, Herbst, A, Hellstrom-Westas, L and Nilsson, A (2007) Human meconium contains significant amounts of alkaline sphingomyelinase, neutral ceramidase, and sphingolipid metabolites. *Pediatr. Res.*, 61, 61–66.

Duivenvoorden, I, Voshol, PJ, Rensen, PC, van Duyvenvoorde, W, Romijn, JA, Emeis, JJ, Havekes, LM and Nieuwenhuizen, WF (2006) Dietary sphingolipids lower plasma cholesterol and triacylglycerol and prevent liver steatosis in APOE*3Leiden mice. *Am. J. Clin. Nutr.*, 84, 312–321.

Eckhardt, ER, Wang, DQ, Donovan, JM and Carey, MC (2002) Dietary sphingomyelin suppresses intestinal cholesterol absorption by decreasing thermodynamic activity of cholesterol monomers. *Gastroenterology*, 122, 948–956.

El Alwani, M, Wu, BX, Obeid, LM and Hannun, YA (2006) Bioactive sphingolipids in the modulation of the inflammatory response. *Pharmacol. Ther.*, 112, 171–183.

Elferink, RP and Groen, AK (1999) The mechanism of biliary secretion and its defects. *Gastroenterol. Clin. North Am.*, 28, 59–74.

Garmy, N, Taieb, N, Yahi, N and Fantini, J (2005) Interaction of cholesterol with sphingosine: physicochemical characterization and impact on intestinal absorption. *J. Lipid Res.*, 46, 36–45.

Graves, EL, Beaulieu, AD and Drackley, JK (2007) Factors affecting the concentration of sphingomyelin in bovine milk. *J. Dairy Sci.*, 90, 706–715.

Heine, W, Wutzke, KD and Radke, M (1993) Sialic acid in breast milk and infant formula food. *Monatsschr. Kinderheilkd*, 141, 946–950.

Hellgren, LI (2001) Occurrence of bioactive sphingolipids in meat and fish products. *Eur J. Lipid Sci. Technol.*, 103, 661–667.

Hellgren, LI (2002) Sphingolipids in the human diet. *Lipid Technology*, 129–133.

Hertervig, E, Nilsson, A, Nyberg, L and Duan, RD (1997) Alkaline sphingomyelinase activity is decreased in human colorectal carcinoma. *Cancer*, 79, 448–453.

Hertervig, E, Nilsson, A, Bjork, J, Hultkrantz, R and Duan, RD (1999) Familial adenomatous polyposis is associated with a marked decrease in alkaline sphingomyelinase activity: a key factor to the unrestrained cell proliferation? *Br. J. Cancer*, 81, 232–236.

Hertervig, E, Nilsson, A, Cheng, Y and Duan, RD (2003) Purified intestinal alkaline sphingomyelinase inhibits proliferation without inducing apoptosis in HT-29 colon carcinoma cells. *J. Cancer Res. Clin. Oncol.*, 129, 577–582.

Idota, T and Kawakami, H (1995) Inhibitory effects of milk gangliosides on the adhesion of *Escherichia coli* to human intestinal carcinoma cells. *Biosci. Biotechnol. Biochem.*, 59, 69–72.

Imaizumi, K, Tomaga, A, Sato, M and Sugano, M (1992) Effects of dietary sphingolipids on levels of serum and liver lipids in rats. *Nutr. Res.*, 12, 543–548.

Inamine, M, Suzui, M, Morioka, T, Kinjo, T, Kaneshiro, T, Sugishita, T, Okada, T and Yoshimi, N (2005) Inhibitory effect of dietary monoglucosylceramide 1-*O*-beta-glucosyl-*N*-2'-hydroxyarachidoyl-4,8-sphingadienine on two different categories of colon preneoplastic lesions induced by 1,2-dimethylhydrazine in F344 rats. *Cancer Sci.*, 96, 876–881.

Kawamori, T, Osta, W, Johnson, KR, Pettus, BJ, Bielawski, J, Tanaka, T, Wargovich, MJ, Reddy, BS, Hannun, YA, Obeid, LM and Zhou, D (2006) Sphingosine kinase 1 is up-regulated in colon carcinogenesis. *Faseb. J.*, 20, 386–388.

Kihara, A, Mitsutake, S, Mizutani, Y and Igarashi, Y (2007) Metabolism and biological functions of two phosphorylated sphingolipids, sphingosine 1-phosphate and ceramide 1-phosphate. *Prog. Lipid Res.*, 46, 126–144.

Kimata, H (2006) Improvement of atopic dermatitis and reduction of skin allergic responses by oral intake of konjac ceramide. *Pediatr. Dermatol.*, 23, 386–389.

Kirby, RJ, Zheng, S, Tso, P, Howles, PN and Hui, DY (2002) Bile salt-stimulated carboxyl ester lipase influences lipoprotein assembly and secretion in intestine: a process mediated via ceramide hydrolysis. *J. Biol. Chem.*, 277, 4104–4109.

Kobayashi, T and Suzuki, K (1981) A taurodeoxycholate-activated galactosylceramidase in the murine intestine. *J. Biol. Chem.*, 256, 1133–1137.

Kobayashi, T, Shimizugawa, T, Osakabe, T, Watanabe, S and Okuyama, H (1997) A long-term feeding of sphingolipids affected the levels of plasma-cholesterol and hepatic triacylglycerol but not tissue phospholipids and sphingolipids. *Nutr. Res.*, 17, 111–114.

Kono, M, Dreier, JL, Ellis, JM, Allende, ML, Kalkofen, DN, Sanders, KM, Bielawski, J, Bielawska, A, Hannun, YA and Proia, RL (2006) Neutral ceramidase encoded by the Asah2 gene is essential for the intestinal degradation of sphingolipids. *J. Biol. Chem.*, 281, 7324–7331.

Lahiri, S and Futerman, AH. (2007) The metabolism and function of sphingolipids and glycosphingolipids. *Cell Mol. Life Sci.* 64, 2270–2784

Li, Z, Basterr, MJ, Hailemariam, TK, Hojjati, MR, Lu, S, Liu, J, Liu, R., Zhou, H and Jiang, XC (2005) The effect of dietary sphingolipids on plasma sphingomyelin metabolism and atherosclerosis. *Biochim. Biophys. Acta*, 1735, 130–134.

Lillienau, J, Cheng, Y, Nilsson, A and Duan, RD (2003) Development of intestinal alkaline sphingomyelinase in rat fetus and newborn rat. *Lipids*, 38, 545–549.

Lingwood, CA, Boyd, B and Nutikka, A (2000) Analysis of interactions between glycosphingolipids and microbial toxins. *Methods Enzymol.*, 312, 459–473.

Lipkin, M, Reddy, B, Newmark, H and Lamprecht, SA (1999) Dietary factors in human colorectal cancer. *Annu. Rev. Nutr.*, 19, 545–586.

Lundgren, P, Nilsson, A and Duan, RD (2001) Distribution and properties of neutral ceramidase activity in rat intestinal tract. *Dig. Dis. Sci.*, 46, 765–772.

Markham, JE, Li, J, Cahoon, EB and Jaworski, JG (2006) Separation and identification of major plant sphingolipid classes from leaves. *J. Biol. Chem.*, 281, 22684–22694.

Moschetta, A, vanBerge-Henegouwen, GP, Portincasa, P, Palasciano, G, Groen, AK and van Erpecum, KJ (2000) Sphingomyelin exhibits greatly enhanced protection compared with egg yolk phosphatidylcholine against detergent bile salts. *J. Lipid Res.*, 41, 916–924.

Moschetta, A, Portincasa, P, van Erpecum, KJ, Debellis, L, Vanberge-Henegouwen, GP and Palasciano, G (2003) Sphingomyelin protects against apoptosis and hyperproliferation induced by deoxycholate: potential implications for colon cancer. *Dig. Dis. Sci.*, 48, 1094–1101.

Moto, M, Okamura, M, Watanabe, T, Kashida, Y and Mitsumori, K (2004) Thirteen-week repeated dose toxicity of rice bran glycosphingolipid in Wistar Hannover (GALAS) rats. *J. Toxicol. Sci.*, 29, 73–80.

Motouri, M, Matsuyama, H, Yamamura, J, Tanaka, M, Aoe, S, Iwanaga, T and Kawakami, H (2003) Milk sphingomyelin accelerates enzymatic and morphological maturation of the intestine in artificially reared rats. *J. Pediatr. Gastroenterol. Nutr.*, 36, 241–247.

Nilsson, A (1968) Metabolism of sphingomyelin in the intestinal tract of the rat. *Biochim. Biophys. Acta*, 164, 575–584.

Nilsson, A (1969a) Metabolism of cerebroside in the intestinal tract of the rat. *Biochim. Biophys. Acta*, 187, 113–121.

Nilsson, A (1969b) The presence of spingomyelin- and ceramide-cleaving enzymes in the small intestinal tract. *Biochim. Biophys. Acta*, 176, 339–347.

Nilsson, A (1970) Conversion of dihydrosphingosine to palmitic acid and palmitaldehyde with cellfree preparations of guinea pig intestinal mucosa. *Acta Chem. Scand.*, 24, 598–602.

Nilsson, A and Duan, RD (2006) Absorption and lipoprotein transport of sphingomyelin. *J. Lipid Res.*, 47, 154–171.

Noh, SK and Koo, SI (2004) Milk sphingomyelin is more effective than egg sphingomyelin in inhibiting intestinal absorption of cholesterol and fat in rats. *J. Nutr.*, 134, 2611–2616.

Nyberg, L, Duan, RD, Axelsson, J and Nilsson, Å (1997) Localization and capacity of sphingomyelin digestion in the rat intestinal tract. *J. Nutr. Biochem.*, 8, 112–118.

Nyberg, L, Farooqi, A, Blackberg, L, Duan, RD, Nilsson, A and Hernell, O (1998) Digestion of ceramide by human milk bile salt-stimulated lipase. *J. Pediatr. Gastroenterol. Nutr.*, 27, 560–567.

Nyberg, L, Duan, R and Nilsson, A (2000) A mutual inhibitory effect on absorption of sphingomyelin and cholesterol. *J. Nutr. Biochem.*, 11, 244–249.

Ogretmen, B and Hannun, YA (2004) Biologically active sphingolipids in cancer pathogenesis and treatment. *Nat. Rev. Cancer*, 4, 604–616.

Ohlsson, L, Palmberg, C, Duan, RD, Olsson, M, Bergman, T and Nilsson, A (2007) Purification and characterization of human intestinal neutral ceramidase *Biochimie*. 89, 950–960.

Olsson, M, Duan, RD, Ohlsson, L and Nilsson, A (2004) Rat intestinal ceramidase: purification, properties, and physiological relevance. *Am. J. Physiol. Gastrointest. Liver Physiol.*, 287, G929–37.

Oskouian, B and Saba, J (2007) Sphingosine-1-phosphate metabolism and intestinal tumorigenesis: lipid signaling strikes again. *Cell Cycle*, 6, 522–527.

Pan, XL and Izumi, T (2000) Variation of the ganglioside compositions of human milk, cow's milk and infant formulas. *Early Hum. Dev.*, 57, 25–31.

Parodi, PW (1997) Cows' milk fat components as potential anticarcinogenic agents. *J. Nutr.*, 127, 1055–1060.

Pettus, BJ, Chalfant, CE. and Hannun, YA (2004) Sphingolipids in inflammation: roles and implications. *Curr. Mol. Med.*, 4, 405–418.

Ramstedt, B and Slotte, JP (2006) Sphingolipids and the formation of sterol-enriched ordered membrane domains. *Biochim. Biophys. Acta*, 1758, 1945–1956.

Rolsma, MD, Kuhlenschmidt, TB, Gelberg, HB and Kuhlenschmidt, MS (1998) Structure and function of a ganglioside receptor for porcine rotavirus. *J. Virol.*, 72, 9079–9091.

Rombaut, R, Camp, JV and Dewettinck, K (2005) Analysis of phospho- and sphingolipids in dairy products by a new HPLC method. *J. Dairy Sci.*, 88, 482–488.

Rombaut, R, van Camp, J and Dewettinck, K (2006) Microfiltration of butter serum upon casein micelle destabilization. *Int. J. Food Sci. Technol.*, 41, 435–443.

Rueda, R, Puente, R, Hueso, P, Maldonado, J and Gil, A (1995) New data on content and distribution of gangliosides in human milk. *Biol. Chem. Hoppe Seyler*, 376, 723–727.
Rueda, R, Maldonado, J, Narbona, E and Gil, A (1998a) Neonatal dietary gangliosides. *Early Hum. Dev.*, 53 Suppl, S135–147.
Rueda, R, Sabatel, JL, Maldonado, J, Molina-Font, JA and Gil, A (1998b) Addition of gangliosides to an adapted milk formula modifies levels of fecal *Escherichia coli* in preterm newborn infants. *J. Pediatr*, 133, 90–94.
Sanchez-Diaz, A, Ruano, MJ, Lorente, F and Hueso, P (1997) A critical analysis of total sialic acid and sialoglycoconjugate contents of bovine milk-based infant formulas. *J. Pediatr Gastroenterol. Nutr.*, 24, 405–410.
Schmelz, EM (2004) Sphingolipids in the chemoprevention of colon cancer. *Front Biosci.*, 9, 2632–2639.
Schmelz, EM, Crall, K., Larocque, R, Dillehay, DL and Merrill, AH Jr (1994) Uptake and metabolism of sphingolipids in isolated intestinal loops of mice. *J. Nutr.*, 124, 702–712.
Schmelz, EM, Bushnev, AS, Dillehay, DL, Liotta, DC and Merrill, AH Jr (1997) Suppression of aberrant colonic crypt foci by synthetic sphingomyelins with saturated or unsaturated sphingoid base backbones. *Nutr. Cancer*, 28, 81–85.
Schmelz, EM, Sullards, MC, Dillehay, DL and Merrill, AH Jr (2000) Colonic cell proliferation and aberrant crypt foci formation are inhibited by dairy glycosphingolipids in 1, 2-dimethylhydrazine-treated CF1 mice. *J. Nutr.*, 130, 522–527.
Schmelz, EM, Roberts, PC, Kustin, EM., Lemonnier, LA, Sullards, MC., Dillehay, DL. and Merrill, AH Jr (2001) Modulation of intracellular beta-catenin localization and intestinal tumorigenesis in vivo and in vitro by sphingolipids. *Cancer Res.*, 61, 6723–6729.
Slotte, JP (1999) Sphingomyelin-cholesterol interactions in biological and model membranes. *Chem. Phys. Lipids*, 102, 13–27.
Smith, DC, Lord, JM, Roberts, LM. and Johannes, L (2004) Glycosphingolipids as toxin receptors. *Semin. Cell. Dev. Biol.*, 15, 397–408.
Sorensen, LK (2006) A liquid chromatography/tandem mass spectrometric approach for the determination of gangliosides GD3 and GM3 in bovine milk and infant formulae. *Rapid Commun. Mass Spectrom.*, 20, 3625–3633.
Sperling, P and Heinz, E (2003) Plant sphingolipids: structural diversity, biosynthesis, first genes and functions. *Biochim. Biophys. Acta*, 1632, 1–15.
Sprong, RC, Hulstein, MF and Van der Meer, R (2001) Bactericidal activities of milk lipids. *Antimicrob. Agents Chemother.*, 45, 1298–1301.
Sugawara, T and Miyazawa, T (1999) Separation and determination of glycolipids from edible plant sources by high-performance liquid chromatography and evaporative light-scattering detection. *Lipids*, 34, 1231–1237.
Sullards, MC, Lynch, DV, Merrill, AH Jr and Adams, J (2000) Structure determination of soybean and wheat glucosylceramides by tandem mass spectrometry. *J. Mass Spectrom.*, 35, 347–353.
Symolon, H, Schmelz, EM, Dillehay, DL and Merrill, AH Jr (2004) Dietary soy sphingolipids suppress tumorigenesis and gene expression in 1,2-dimethylhydrazine-treated CF1 mice and ApcMin/+ mice. *J. Nutr.*, 134, 1157–1161.
Tsuji, K, Mitsutake, S, Ishikawa, J, Takagi, Y, Akiyama, M, Shimizu, H, Tomiyama, T and Igarashi, Y (2006) Dietary glucosylceramide improves skin barrier function in hairless mice. *J. Dermatol. Sci.*, 44, 101–107.
van Veldhoven, PP and Mannaerts, GP (1993) Sphingosine-phosphate lyase. *Adv. Lipid Res.*, 26, 69–98.
Vesper, H, Schmelz, EM, Nikolova-Karakashian, MN, Dillehay, DL, Lynch, DV and Merrill,

AH Jr (1999) Sphingolipids in food and the emerging importance of sphingolipids to nutrition. *J. Nutr.*, 129, 1239–1250.

Wang, B and Brand-Miller, J (2003) The role and potential of sialic acid in human nutrition. *Eur. J. Clin. Nutr.*, 57, 1351–1369.

Wang, B, Petocz, P and Miller, JB (2003) Relationship of sialic acid and fatty acid composition of brain gangliosides: breast-fed vs formula-fed infant. *Asia Pac. J. Clin. Nutr.*, 12, Suppl, S43.

Wang, L, Wang, T and Fehr, WR (2006) Effect of seed development stage on sphingolipid and phospholipid contents in soybean seeds. *J. Agric. Food Chem.*, 54, 7812–7816.

Wehrmuller, K (2007) Occurrence and biological properties of sphingolipids – a review. *Curr. Nutr. Food Sci.*, 3, 161–173.

Wu, J, Cheng, Y, Nilsson, A and Duan, RD (2004) Identification of one exon deletion of intestinal alkaline sphingomyelinase in colon cancer HT-29 cells and a differentiation-related expression of the wild-type enzyme in Caco-2 cells. *Carcinogenesis*, 25, 1327–1333.

Zeisel, SH, Char, D and Sheard, NF (1986) Choline, phosphatidylcholine and sphingomyelin in human and bovine milk and infant formulas. *J. Nutr.*, 116, 50–58.

Zheng, W, Kollmeyer, J, Symolon, H, Momin, A, Munter, E, Wang, E, Kelly, S, Allegood, JC, Liu, Y, Peng, Q, Ramaraju, H, Sullards, MC, Cabot, M and Merrill, AH, Jr (2006) Ceramides and other bioactive sphingolipid backbones in health and disease: lipidomic analysis, metabolism and roles in membrane structure, dynamics, signaling and autophagy. *Biochim. Biophys. Acta*, 1758, 1864–1884.

Abbreviations

ACF	aberrant colonic crypt foci
ABA	abscisic acid
AOM	azoxymethane
APC	adenomatous polyposis coli
BMP	bis(monoacylglycerol) phosphate
BSSL	bile salt stimulated lipase
CDP	cytidine diphosphate
CLA	conjugated linolenic acid
CMC	critical micelle concentration
CoA	coenzyme A
CPI	ceramide phosphatidyl inositol
CPP	critical packing parameter
DGDG	digalactosyldiacylglycerol
DGG	digalactosylglycerol
DGMG	digalactosylmonoacylglycerol
DGPP	diacylglycerolpyrophosphate
DHA	docosahexaenoic acid
DLPA	dilauroylphosphatidic acid
DLPC	dilauroylphosphatidylcholine
DLPE	dilauroylphosphatidylethanolamine
DLPG	dilauroylphosphatidylglycerol
DLPS	dilauroylphosphatidylserine
DMH	dimethylhydrazine
DMPA	dimyristoylphosphatidic acid
DMPC	dimyristoylphosphatidylcholine
DMPE	dimyristoylphosphatidylethanolamine
DMPG	dimyristoylphosphatidylglycerol
DMPS	dimyristoylphosphatidylserine
DOPG	dioleoylphosphatidylglycerol
DOPA	dioleoylphosphatidic acid
DOPC	dioleoylphosphatidylcholine
DOPE	dioleoylphosphatidylethanolamine
DPG	diphosphatidyl glycerol
DPPA	dipalmitoylphosphatidic acid

DPPC	dipalmitoylphosphatidylcholine
DPPE	dipalmitoylphosphatidylethanolamine
DPPG	dipalmitoylphosphatidylglycerol
DPPS	dipalmitoylphosphatidylserine
DSPA	distearoylphosphatidic acid
DSPC	distearoylphosphatidylcholine
DSPE	distearoylphosphatidylethanolamine
DSPG	distearoylphosphatidylglycerol
DSPS	distearoylphosphatidylserine
EPA	eicosapentaenoic acid
FAP	familial adenomatous polyposis
FDA	Food and Drug Administration (US)
FFA	free fatty acid
GPC	glycerophosphocholine
G-Tases	glycosyltransferases
HDL	high density lipoprotein
HLB	hydrophilic–lipophilic balance
LC-PUFA	long-chain polyunsaturated fatty acid
LPC	lysophosphatidylcholine
LPG	lysophosphatidylglycerol
MCM	medium-chain monoacylglycerol
MCT	medium-chain triacylglycerol
MGDG	monogalactosyl diacylglycerol
MGMG	monogalactosyl monoacylglycerol
MLV	multilamellar liposome
NAPE	*N*-acyl phosphatidylethanolamine
PA	phosphatidic acid
PBR	packed-bed reactor
PC	phosphatidylcholine
PE	phosphatidylethanolamine
PG	phosphatidylglycerol
PI	phosphatidylinositol
PIP_2	phosphatidylinositol 4,5-bisphosphate
PL	phospholipid
PLA_1	phospholipase A_1
PLA_2	phospholipase A_2
PLC	phospholipase C
PLD	phospholipase D
POPC	1-palmitoyl-2-oleyl phosphatidylcholine

POPG	1-palmitoyl-2-oleyl phosphatidylglycerol
PS	phosphatidylserine
PUFA	polyunsaturated fatty acid
S1P	sphingosine-1-phosphate
SAM	*S*-adenosylmethionine
SM	sphingomyelin
SUV	small unilamellar liposome
VLDL	very high density lipoprotein

Index

FSC
www.fsc.org
MIX
Papier | Fördert gute Waldnutzung
FSC® C083411

Kolibri 360 GmbH
Ferdinand-Jühlke-Straße 7
99095 Erfurt, Deutschland
produktsicherheit@kolibri360.de